Developmental Human Behavior Genetics

Developmental Human Behavior Genetics

Nature-Nurture Redefined

Edited by

K. Warner Schaie
University of Southern
California

V. Elving Anderson
University of Minnesota

Gerald E. McClearn
University of Colorado

John Money
The Johns Hopkins University

Lexington Books
D.C. Heath and Company
Lexington, Massachusetts
Toronto   London

Library of Congress Cataloging in Publication Data

Main entry under title:

Developmental human behavior genetics.

   Includes index.
   1. Behavior genetics. 2. Developmental psychology. 3. Nature and nurture. 4. Human genetics. I. Schaie, Klaus Warner, 1928-
[DNLM: 1. Genetics, Behavioral—Congresses. 2. Psychology—Congresses. BF713 D489 1974]
QH457.D48   155.7   75-733
ISBN 0-669-99515-0

Copyright © 1975 by D.C. Heath and Company

All rights reserved. No part of this publication may be reproduced or transmitted in any form or by any means, electronic or mechanical, including photocopy, recording, or any information storage or retrieval system, without permission in writing from the publisher.

Published simultaneously in Canada

Printed in the United States of America

International Standard Book Number: 0-669-99515-0

Library of Congress Catalog Card Number: 75-733

Contents

	Preface	vii
Chapter 1	Introduction: Behavioral Genetics and Developmental Psychology, *Gerald E. McClearn*	1
Chapter 2	Gene-environment Interaction in Human Behavioral Development, *Norman D. Henderson*	5
	Commentary I, *L. Erlenmeyer-Kimling*	25
	Commentary II, *Merrill F. Elias*	33
Chapter 3	Empirical Methods in Quantitative Human Behavior Genetics, *John C. Loehlin*	41
	Commentary, *Steven G. Vandenberg*	55
Chapter 4	Populations for the Study of Behavior Traits, *G. Ainsworth Harrison*	65
	Commentary I, *Sandra Scarr-Salapatek*	77
	Commentary II, *Lissy F. Jarvik*	85
Chapter 5	Genetic Mechanisms in Human Behavioral Development, *Gilbert S. Omenn*	93
	Commentary I, *V. Elving Anderson*	113
	Commentary II, *Barton Childs*	119
Chapter 6	Quantitative Genetic Perspectives: Implications for Human Development, *L.L. Cavalli-Sforza*	123
	Commentary I, *I. Michael Lerner*	139
	Commentary II, *John C. DeFries*	145
Chapter 7	Counseling in Genetics and Applied Behavior Genetics, *John Money*	151
	Commentary I, *Sheldon C. Reed*	171
	Commentary II, *Robert F. Murray*	173
Chapter 8	Ethical Issues in Human Behavior Genetics: Civil Rights, Informed Consent and Ethics of Intervention, *Arthur Falek*	179

	Commentary I, *Bruce K. Eckland*	197
	Commentary II, *James F. Crow*	201
Chapter 9	**Research Strategy in Developmental Human Behavior Genetics,** *K. Warner Schaie*	205
Chapter 10	**Possible Directions for Developmental Human Behavior Genetics,** *Irving I. Gottesman*	221
	Glossary	229
	Author Index	237
	Subject Index	241
	Invited Workshop Participants	245
	Members of the Developmental Behavioral Sciences Study Section, National Institutes of Health	247
	About the Editors	249

Preface

Questions raised by human behavior geneticists are of serious importance to the quality of human life. Nevertheless, psychologists, sociologists, and other social scientists have concentrated on environmental issues to the point that the interaction of genotype and environment has been seriously neglected. Reluctance to introduce behavior genetics into social science curricula and research programs may result from an extreme reaction by social scientists to the misuse of genetics in the 1930s as well as the recent controversies regarding the meaning of heritability estimates for public policy issues such as compensatory education.

One of the most noteworthy areas of neglect within the area of behavior genetics has been the investigation of the relationship of genotype and environment as they affect development. But this is the particular facet of behavior genetics that would be of greatest interest to behavioral and social scientists since it is obvious that development must occur as a function of the constantly changing process of interactions between genotypes and environment.

Animal behavior genetics emerged late on the scene in the behavioral sciences and is only now becoming firmly established. At the infrahuman level there has also been a lack of substantial explorations regarding interaction of genotype, environment, and development, but there have been a few developmental studies and a recognition that more are needed. At the human level behavior genetic investigations have concentrated even more on static issues than would be true for the investigation or animal models. There are quite a number of investigations conducted at various stages of development, but few studies have dealt with changes in behavior either for a brief segment of life or over the lifespan of individuals. This state of affairs may in part reflect the fact that developmental psychologists are usually not well versed in genetics and that geneticists have traditionally searched for gene effects that are invariant and more closely related to gene action than is behavior.

During the past few years, however, there has been an increasing interest in a more developmentally oriented approach to human behavior genetics. As chairman of the Developmental Behavioral Sciences Study Section in the Division of Research Grants at the National Institutes of Health (NIH) I became aware of the increasing number of requests for support for research in this area. Careful consideration of these requests posed a variety of serious substantive and methodological issues. Not only is human behavior genetics research expensive, but the issues raised have important ideological and political consequences. In addition, few social and behavioral scientists, other than the relatively small number of researchers who are submitting proposals in this area, are qualified by training to evaluate properly the merits of behavior genetics projects, either from a scientific standpoint or with regard to possible political and ideological implications.

Because of these considerations I proposed to the NIH Division of Research Grants that the level of informed review of proposals in the area of human behavior might be improved and the current state of the art be given better definition by conducting a workshop-conference that would include members of our study section, NIH personnel involved in the management of grants and contracts in the area of inquiry, and experts in behavior genetics. With the strong support of my colleagues and co-editors, V. Elving Anderson, Gerald E. McClearn, and John Money, and the executive secretary of the Developmental Behavioral Sciences Study Section, Bertie H. Woolf, I was able to persuade the Division of Research Grants to approve a two-day workshop to be held at the National Institutes of Health in Bethesda, April 16 and 17, 1974. Financial support for the workshop-conference was provided by the Scientific Evaluation (Chairman's) Grant #3 R09 GM 16639-02 with additional support from the National Institute of Child Health and Human Development.

In preparation for the workshop-conference seven position papers were commissioned covering selected issues relating to problem definition, research methodology, and public policy implications. Each position paper was then, in advance of the workshop, reviewed by two other experts who prepared formal comments. At the workshop-conference, presentation of the position papers and formal responses was followed by critical discussions in which the invited experts and study section members interacted vigorously.

This book includes edited revisions of the seven position papers and thirteen responses. It also includes Gerald McClearn's introductory remarks, which set the stage for the workshop discussions, and Irving Gottesman's closing statement, which points to some of the future directions. During the workshop relatively little emphasis was given to attempts to link developmental and behavior genetic research strategies, and I therefore elected to add some notes on research strategy in developmental human behavior genetics (Chapter 9). Since the purpose of the workshop-conference and this book is to allow human behavior geneticists to address a broad audience of social and behavioral scientists it also seemed useful to add a glossary of terms.

The editors wish to express their gratitude to the Division of Research Grants and the National Institute of Child Health and Human Development for their financial support; to Dr. Bertie H. Woolf for his capable administrative and conceptual support in mounting the conference; to Mrs. Agnes Abrams for her help with the physical conference arrangements; to Dr.

Donna Cohen for preparing the glossary; to my wife, Dr. Joyce P. Schaie for preparing the indexes; and to my secretary Laura Bielefelt for coordinating the many clerical activities relating to the conference and manuscript editing tasks.

K. WARNER SCHAIE

**Developmental Human
Behavior Genetics**

1

Introduction: Behavioral Genetics and Developmental Psychology

Gerald E. McClearn
University of Colorado

Much of the research in behavioral genetics, particularly the early work, has been concerned with demonstrating that the basic Mendelian principles and their extensions in quantitative genetics are applicable to behavior. It seems fair to say that this demonstration has been convincing. In practically all cases where genetic influence on a particular behavior has been sought, it has been found. This ubiquity of hereditary influence on behavior has important implications for psychology, perhaps the most important being that a revised perspective on individual differences is appropriate.

Briefly, the Mendelian laws describe the workings of meiosis, a process of genetic sampling from within the individual in the formation of his or her gametes, and of sexual reproduction, in which two gametes, egg and sperm, unite to begin another individual. The crucial point is that this mechanism systematically generates genetic variability. It has been a biological necessity that this be so, because intraspecific variability is the sine qua non of the evolutionary process. Now, this system works so well that it is very nearly *incapable* of generating the same genetic constitution in any two members of a sexually reproducing species. (The exception, and it is trivial in terms of population dynamics, is the case of identical twins, and other multiple identical births). Thus, if all of the sets of environmental influences that have been identified—nutrition, rearing mode, peer groups, education, and the rest—could be equally applied to all, there would still be differences among us. Furthermore, the environmental differences that exist in the real world and the genetic differences do not necessarily simply summate. Individuals of some genotypes may respond differently to a particular environment than individuals of other genotypes. The appropriate model, then, is one that acknowledges that *both* genetics and environment are essential for the development of the organism and thus for any of its traits. Such a model views variability as attributable to main effects of genes, main effects of environment, and interaction terms. This perspective has direct and immediate relevance to those parts of psychology that are in the individual differences business—psychometrics and differential psychology. It offers a whole realm of explanatory principles for the origin of individuality that has not been available to many of the workers in those fields.

There are also implications for the rest of psychology. When individual differences have not been under direct scrutiny they have often been swept into an error term, and it has been easy to come to view variability as really being only error—mistakes of control or measurement—that would approach the vanishing point if only a tight enough experimental design and reliable enough measures could be discovered. Well, it is not so, and the mind boggles at the thought of how many of the cherished findings of animal behavioral research, for example, are idiosyncratic to the particular group of animals studied. The demonstrations of differences among carefully controlled inbred and selected strains, not only in magnitude but in direction of effect of an independent variable, hint at the amount of confusion that might have been introduced into our literature by the use of casually chosen and haphazardly maintained stocks. Serving as a bridge to communicate the genetic perspective on individuality to the behavioral sciences is, thus, one of behavioral genetics' important functions.

Behavioral genetics, in addition to attempting to come to grips with individuality at the quantitative level, is concerned with the mechanisms through which the genes exert their influences on behavioral traits. In this type of study genes are manipulated as independent variables by the choice of inbred strains, selective breeding, utilization of single gene mutations, etc. The dependent variables may include behavior and the enzyme systems, neural structures or functioning, or endocrine activity—all thought to be part of the causal pathway. This latter area is, of course, the domain of physiological psychology, or biopsychology. Biopsychology is itself a bridge between the biological and the behavioral sciences, but it has traditionally emphasized physiology and has had little to do with genetics on the biological side. In this case, then, behavioral genetics is a bridge to a bridge.

In dealing with the principles of population genetics and applying them to behavior, behavioral geneticists are concerned with both the forces that maintain or alter across generations the frequencies of genes that influence behavior and also the effects of behavior on the gene flow from one generation to the next. In the short run the phenomena of interest include such matters as transgenerational trends in intelligence, maintenance in the population at high frequency of the gene or genes influencing schizophrenia, class structure, effects of programs of genetic counselling on the incidence of various types of mental retardation, and so on. In the longer run the concerns relate to behavioral evolution, and behavioral genetics is to be seen vis-à-vis comparative psychology and ethology.

Behavioral genetics consequently means many things to many disciplines. But above all it is a handy multilevel bridge between the biological and behavioral sciences, and more traffic on the bridge will probably be a very useful thing.

Our purpose in the workshop that led to this book was to explore the nature of the link that behavioral genetics provides between developmental biology and developmental psychology. One of the perennial and key problems in biology has been the astounding developmental process whereby a single fertilized egg becomes a functioning, integrated organism. One of the most intriguing subsets of the problems of developmental biology has been developmental genetics—how the genes influence ontogeny. Research on microorganisms, showing that genes can be turned on and off, have offered exciting models for developmental processes. In more complex organisms it has even been possible to relate observable chromosome changes to developmental events. These and similar researches are signs of things to come. It requires no special acuity of perception to predict that developmental biology will become an increasingly lively and central research area in the years just ahead. In the meantime developmental psychology has become increasingly active and robust. It is not a foregone conclusion that these two areas will make common cause, but it will be a great pity if they do not. Perhaps this book will augment the prospect that they will.

2

Gene-environment Interaction in Human Behavioral Development

Norman D. Henderson
Oberlin College

As an investigator working largely with laboratory animals I admire greatly the tenacity of any investigator willing to undertake a large-scale empirical study involving in part an examination of genetic influences on behavior in human populations. Unforeseen complications in animal research on behavioral genetics are often technical or practical ones similar to those in most areas of animal psychology. While there are some difficulties with models, assumptions, and forms of analyses in animal research, the fact that experimental control and manipulation can be applied to a far greater extent than in human studies reduces complications enormously. Investigators undertaking a large-scale study of a human population, in spite of the most careful preparations, can fully expect a series of post hoc criticisms about their assumptions, sample selection and size, and general forms of analyses and their interpretations. Indeed, for every hardy soul undertaking such an experiment, I expect that there are ten critics, often with legitimate and constructive ideas who are ready to point out flaws in the research program. Further, for each of these critics, who provide a useful and necessary function in the field, there are probably ten additional persons who will use such criticism as a basis for misinterpreting or altogether dismissing the entire research effort. I think it is fair to say, for example, that virtually no study presently available on the genetic basis of intelligence cannot be criticized on several grounds. Nevertheless, one cannot help but be impressed by the apparent consistency that seems to occur in the overall picture relating degree of kinship and similarity on various IQ measures, as summarized for example in Erlenmeyer-Kimling and Jarviks (1963) summary of some of this work, although even summaries such as this do not escape criticism.

Perhaps the key difference in animal and human genetic research with respect to control of variables is that in the former an investigator can eliminate, control for, or examine systematically assortive mating, most postnatal sources of genotype by environment interactions, and to a lesser degree reduce correlations that may exist between genotypes and environments. On the other hand, each of these problems may exist in human populations, and if they do, they confound estimates of genetic and environmental influences on behavior.

Before discussing gene-environment interactions in human behavioral development I would like to describe very briefly the results of some genetic experiments on mice reared under different environmental conditions that were done at our laboratory in Oberlin. Hopefully these simple examples will serve to illustrate certain issues and outcomes that must be considered possible in human behavioral genetics research. With respect to the understanding of human behavior these examples can be illustrative only. With the possible exception of some rather limited work dealing with physiological and biochemical processes, and some aspects of social interaction, temperament, and basic conditioning processes, the use of animal experiments as supporting evidence for theories concerning complex human abilities is a risky business with a rather unsuccessful history.

Much speculation concerning early environmental and genetic influences on animal learning and intelligence based on studies of the late nineteen fifties and early nineteen sixties has been retracted or radically revised to the point that little of this work cited in early reviews such as Hunts' *Intelligence and Experience* (1961) can be considered particularly relevant to the question of human intellectual capacity. Furthermore, it should be obvious that while certain genetic mechanisms or patterns may emerge in these animal studies, there is little reason to expect that similar patterns will occur in human research, although some factors, because they are related to fitness or overall selective advantage for an organism might be similar. Thus, the rather consistent finding in research with rats and mice that various learning behaviors exhibit genetic dominance in the direction of better performance suggests that behavioral plasticity, or learning ability, has some adaptive significance for these species. On intuitive grounds it seems reasonable to expect that the ability to learn from experience would have an adaptive function for all species, and insofar as some of our aptitude and intelligence measures reflect learning ability we should find some degree of directional dominance in these traits as we do in lower animals.

Let us look at a few studies that attempt to examine gene × environment (GE) interactions in mice. The breeding design used is a diallel cross, which consists of crossing a number of inbred lines into all possible N^2 combinations. A number of methods of analyzing such designs are now available that allow different estimates of several genetic parameters, including additive and several sources on nonadditive genetic variance as well as maternal effects and sex-linked effects. If such a design is used on a population reared in a more or less uniform environment, such as in typical laboratory cages, the results one obtains must be considered relevant only for subjects reared under that rather narrow range of environmental variation. Within this environment one can examine one type of GE interaction by examining the relative variability of each genotype used in the design, giving some picture of the interaction between genotypes and the microenvironment. This entire genetic plan of N^2 genotypes can be replicated in a radically different environment that may involve a different level of nutrition, social interaction, environmental complexity, or another major variable or variables of interest. Genotypes × macroenvironmental interactions will then be represented as significant interaction terms in the expanded analysis of variance including results from both replications, and will lead to different genetic architecture being found in the different environments.

In our research, for example, in addition to carrying out the diallel cross analysis of a variety of behaviors of mice reared in standard laboratory cages we also reared an equivalent set of genotypes in cages allowing considerably more exploration and interaction with stimulus objects. After six weeks of living in one of these environments, mice were tested in a simple food seeking task. The hungry mouse was placed on the floor of a large enclosure and its time required to find its way to the food basket located on the upper portion of one wall was measured. An analysis of the performance of mice reared in our standard lab cages suggested that this particular behavior was influenced to only a small degree by genetic factors—about 10 percent. On the other hand, the groups of animals that had spent the first six weeks of their lives in the more complex cages not only performed considerably better on the average in this task, but there was also much greater variability between genotypes, resulting in a heritability of 40 percent (Henderson 1970a). An examination of mean performance for various genetic combinations reared in the two environments demonstrates rather clearly what happened (Figure 2-1). In essence, some genotypes showed virtually no difference in performance as a result of rearing under these two different conditions, while other lines showed a substantial increase in performance by their group reared in the more complex environment. From the figure one can see the relatively narrow range of scores among different hybrid combinations reared in standard cages as compared with that of hybrids reared in enriched cages.

This result probably represents the most extensive animal demonstration of the "reaction range" concept described by Gottesman (1963) and subsequently referred to by a number of investigators dealing with human behavior. In essence this model suggests that as the favorableness of an environment increases, the degree of genetic expression of a character also increases, both for the phenotypic variance and the heritability of the character in question. This model has great appeal, for it is intuitively sensible to think that an enriched or more favorable environment would allow genetic effects to be expressed more fully than they would be in a situation that might be described by some as environmental impoverishment. Also, while improvement is differential among different genotypes,

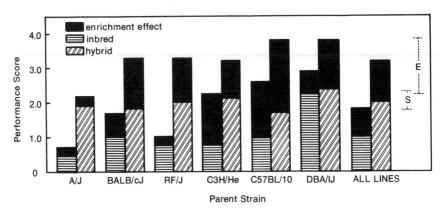

Source: Norman D. Henderson, "Genetic Influence on the Behavior of Mice Can Be Obscured by Laboratory Rearing," *Journal of Comparative & Physiological Psychology*, vol. 72, p. 507. Copyright 1970 by the American Psychological Association. Reprinted by permission.

Figure 2-1. Mean Performance for Various Genetic Combinations Reared in Two Environments.

the performance of all groups increases as a function of environmental improvement.

While in this model and in our example of the mice finding food heritability was greater in the more enriched environment, it is possible to have other forms of GE interaction that have more modest effects on overall heritability but nevertheless indicate a pronounced difference in the genetic architecture. For example, brain weights of mice reared in the two environments just described were also examined (Henderson 1970b, 1973). Among subjects reared in standard cages genetic effects were largely additive, while among mice reared in the more complex environment, hybrids had a significantly higher brain weight than their inbred parents, indicating directional dominance toward larger brain size. While the total variance accounted for by genetic factors was not particularly different in these two situations, the nature of the genetic influences differed in such a way that our conclusions about the possible evolutionary significance of brain size would have differed considerably if we had only studied standard laboratory reared mice.

While the reaction range model of GE interaction has received more attention than any other, it is only one of a rather large family of possible interactions. The reaction range concept suggests that a shift to a more favorable environment increases individual differences through an increase in genetic variation. It is also quite possible for phenotypic variance to be increased in a more favorable environment through increases in microenvironmental influences, with genetic variation remaining constant. Alternatively, phenotypic variance could increase in a new environment through both increased genetic and environmental variance. Heritability would of course differ in these three cases, increasing in the first, decreasing in the second, and not changing in the third situation. In addition to these possible shifts in phenotypic variance in different environments, there can also be shifts in overall population means. A new environment could produce an increase in overall mean performance, as in our mouse food-seeking example, a decrease in means, or no overall change in the population performance of the character being measured.

One must also consider cases where either genetic or environmental variance would decrease in a new situation. There are then 3 x 3 x 3, or twenty-seven possible combinations of differences (or lack of them) in means, environmental variation and genetic variation that could exist between two different environments. One of these represents a no change condition and thirteen are symmetrical with the remaining thirteen. Therefore, while most attention has been paid to those situations where a change in environment results in increasing mean performance and greater genetic variability, thirteen additional GE interaction patterns exist. Although some of these may never be demonstrated empirically, a number have already been reported in animal research. In fact, when one looks closely at the animal studies in which experimenters systematically varied both genotypes and environmental treatment, GE interactions tend to occur more often than not (Erlenmeyer-Kimling 1972). While we must take this seriously with respect to generalizations of results in animal studies, the nature of the treatment and the genetic material used in these studies is generally such as to make them not particularly useful in anticipating the frequency or magnitude of GE interactions that one would normally expect to find in studying human behavior.

Just what can we say about the expression of genetic characters on behavior in widely differing circumstances in human populations? The answer is, not much, for the simple reason that this strategy of replicating a genetic design in different cultures or other differing environmental conditions has been carried out so rarely that we still have very little information. One study that comes close to this type of design is that of Scarr-Salapatek (1971), who reported lower heritability estimates among black children in Philadelphia than among whites. Because differences in the heritability estimates of these two populations were based on differences in difference scores among twins, the reliability of this result has been questioned (Allen and Pettigrew 1973; Erlenmeyer-Kimling and Stern 1973). Aside from other difficulties, Scarr-Salapatek ran into problems of sample size in attempting to compare heritability estimates across populations using standard procedures with unseparated monozygotic and dizygotic twins. Her total of nearly 1,000 pairs was just not sufficient for these kinds of comparisons. Even within a given environmental or age range, obtaining adequate sample sizes poses great difficulties for investigators interested in obtaining reliable estimates of various genetic parameters. It is not surprising therefore that investigators are reluctant or unable to undertake adequate studies in several different populations, in spite of recent developments that have extended analytic procedures to include more than one strategically chosen group. Most data that we presently have therefore is based on work done within a single but usually broad ranged population category, most typically a middle class Caucasian sample from the United States or Europe.

Thus far in the literature twin studies predominate. At the minimal level these involve MZ-DZ comparisons where the twins are reared together. More extensive studies include sibs reared together or reared apart and unrelated children reared together. Finally, some investigators have been fortunate enough to obtain a moderately large sample of monozygotic twins reared apart for inclusion in their analyses. The correlation between monozygotic twins reared apart examined in conjunction with estimates of parent offspring correlations and the correlation between parents scores, allows a rather complete decomposition of genotypic variance into additive genetic, dominance and assortive mating effects, as well as environmental effects. While the validity of some of the data involved are in question, a rather complete example of this can be found in Burt (1972) with respect to the analysis of intelligence scores in his London sample.

Even at this elaborate level, however, the classical approach to correlations still depend on the assumption that phenotypic variance between individuals is a simple sum of genetic and environmental variances and that both GE interactions and the covariance or correlation between genotype and environment are negligible. The consequences of these assumptions are important, since the presence of GE interactions will always increase total phenotypic variance. On the other hand if genotypic and environmental contributions are correlated, variance can be either increased or decreased. Classical twin methodology recognizes these limitations but is unable to detect or correct for effects of these sources of variation and thus depends on indirect evidence that their effects are not substantial. This assumption, of course, has been challenged by several critics who point to evidence from numerous animal husbandry studies or derive hypothetical examples for human behavioral characters that seem quite plausible.

What was needed here were some extensions of the classical twin approaches to test these assumptions and if necessary provide corrections. A major step in this direction is taken by Cattell (1953), when he introduced a multiple variance analysis method for psychological genetics. His brief outline was expanded and the technique later renamed multiple abstract variance analyses or (MAVA) by Cattell (1960). A few years later Loehlin (1965) corrected some inconsistencies in some of Cattell's equations. In 1970 Jinks and Fulker (1970) added some preliminary tests of basic assumptions underlying simple genetic environmental models that can be used with either MAVA, or the biometrical approach to analyzing gene actions; Eaves (1972a, 1972b) has provided useful information concerning sample sizes and multivariate extensions of some of the biometrical genetic methods. Recently Cattell (1973) extended his multiple abstract variance analysis method into what is now called the comparative MAVA method involving comparisons of variances and correlations of different populations such as different cultural, racial, or age groups, or between populations with different degrees of inbreeding. While only briefly outlined at this point, Cattell, in attempting to examine developmental curves, proposes to split further the environmental components of such curves into average environmental and special epoch components using the methods developed by Schaie (1965) and later Baltes (Baltes et al. 1970) that involve comparisons of cursive longitudinal and cross sectional data (Cattell 1973). Cattell would also apply MAVA to environmental feature measures and has suggested a general model for causal connections among genetic and environmental features and genetic and environmental components in the individuals (also see Chapter 9).

These admirable and ambitious extensions of the MAVA method undoubtedly represent an extremely powerful set of tools for examining genotype, age, and environmental interactions in a much broader sense as presently done within single population samples. Unfortunately, the required size of these multiple population analyses using MAVA are simply staggering. Early estimates of required sample size for the full MAVA analyses within one population quickly jumped from 500 to 5,000 and using Eaves (1972a) estimates, even this number appears to be low unless the experiment is designed to be as near as possible to the optimal with respect to the proportions of different kinship groups included.

The difficulty of obtaining this number of subjects in each of several different populations is obvious. While the full design requires a large number of familiar groups yielding sixteen experimentally determinable variances, a limited resource design is also proposed involving ten such variances. This reduction is not without cost, however, as one must then assume that environmental variance for identical and fraternal twins raised together is identical and one must omit differentiation of some of the genotype environment correlations obtained in the full model.

The limited resource model does eliminate the need for identical twins reared apart, clearly the most difficult group to obtain, but also eliminates easily obtained sibs reared together, and half sib data that are useful for narrow heritability estimates. Any reduction in the full design always involves compromises, and in this light perhaps the more radical compromise suggested by Jinks and Fulker (1970) should be used, namely by MZ and DZ twins reared together and sibs reared apart. This combination probably represents the most economical composite of groups to locate, which may then permit larger samples or multiple environmental levels to be studied. This latter design precludes the use of one of the two tests for genotype environmental interaction—that involving within-family environmental differences. If this aspect of genotype environmental interaction proves negligible in most behavioral characters this design has clear economic advantages. Other possibilities, such as offspring of identical twins, also deserve further exploration.

At present, what can we say about the magnitude of genotype interactions in various studies in which these can be examined? A simple test used by Jinks and Fulker (1970) for this purpose involves correlating sums and differences of monozygotic twin scores. If all twin pairs are affected to the same extent by the environmental influences within the family, then pair differences will be equal across twin sets, within sampling error. On the other hand, if twins in some families are exposed to different environmental influences than other families, or if twins react differently in these families, pair differences will not be equal. In addition, the sum of twin scores will differ if twins belonging to different families have different family environments, different genotypes, or both. An interaction between genotypes and within-family environment will thus be reflected in a correlation between the twin sums and twin differences over the N families. For twins that have been reared apart one can test for an interaction between genotypes and environmental differences between families in the same way. Jinks and Fulker have applied these techniques to previously generated data from a number of studies in the personality and intelligence domain primarily using Shields' (1962) data, which is readily amenable to analysis for GE interactions. In examining neuroticism as measured by a modified version of the Maudsley Personality Inventory, they found no evidence of gene environment interaction, nor was there evidence of correlated environments with genotype. In essence, a simple genetic model was adequate to explain the data consisting only of additive gene action and some assortive mating. Common family environments were also not important.

On an introversion-extroversion scale, however, one might expect some GE interaction, since introverts are generally more conditionable than extroverts. This was the case in Shields' data, at least with respect to genotype × within-family environmental interaction. Introverts were in fact more susceptible to environmental influences than extrovert genotypes. On the other hand, while this GE interaction was significant, the results of a genetic analysis using a simple no-interaction model and a model including this interaction were quite similar, since the G × E effect accounted for only 6 percent of the total phenotypic variation.

Jinks and Fulker also analyzed Shields' twin data on two tests of intellectual ability, a vocabulary test and a Dominoes intelligence test involving perceptual-cognitive tasks similar to those of Raven Progressive Matrices Test. The analysis of the vocabulary test indicated that a small degree of G × E interaction, again involving within-family environment was present. Individuals with lowest scores appeared more influenced by their environment than those with higher scores. Again, however, only a small percentage, in this case 8 percent, of the phenotypic variance could be attributed to an environmental interaction with genotype. The nonverbal Dominoes test, which would probably be regarded as a "culture fair" intelligence test, indicated no significant GE interaction, either between or within families. Thus, in Shields' data it was found that on the personality measure and the aptitude measure on which one would expect strong environmental influences, low-magnitude genotype × within-family environmental interactions were found. It should be added that genotype environment *correlations* did not exist on any of these measures. On one additional small sample of MZ twins tested on the Otis intelligence scale no G × E interactions were found by Jinks and Fulker. More recently the sum-difference procedure was used by Jensen (1970) in the analysis of IQs of identical twins reared apart. His work represented a reanalysis of combined earlier data, including some of Shields' data mentioned above with the performance on the verbal and perceptual tests combined. The total sample consisted of 122 pairs of twins reared apart. The correlation of sums and differences between twin pairs was nonsignificant, suggesting no GE interaction for IQ as defined by these tests. From these various analyses it appears that GE interactions are not of great consequence in many of the typical studies on human abilities. While a considerable amount of additional data will probably be subject to this type of analysis, I see little reason to expect that the situation will change very much from that which has been reported in these early data returns. Why should this be?

First, of course, one might argue that the environmental conditions for the populations studied above were relatively homogeneous. While the range of environmental conditions under which homo sapiens develop is certainly far greater than that included in these studies, it is unreasonable to regard this combined group of Caucasian populations from the United States, England, and Denmark as being subjected only to microenvironmental variation as in the case of laboratory animals. The reasons for the relatively negligible GE interactions for these populations might be more appropriately attributed to a nature of the measurements themselves. Most good measuring instruments or tests of personality or intelligence are generally constructed in such a way as to maximize test-retest reliability. In other words, most test makers have gone to great efforts to choose items that are relatively buffered from the short-term influence of minor environmental fluctuations that individuals are subject to. This conscious effort to provide stability of measures has likely resulted in minimizing short-term environmental or GE interaction effects. As a result, analysis of data from such tests is less likely to indicate a complex mode of inheritance including genotype-environment interactions than one might find on behavioral measures other than these standardized tests.

As previously mentioned, these reanalyses of earlier data also indicated no evidence for the presence of correlations between genotype and environment. This is fortunate, since one form of genotype-environment correlation, that dealing with within-family hereditary and environmental effects *cannot be estimated* with present methods, making the investigation of gene action very difficult when significant correlations of this type are found. The presence of such correlations is intuitively very reasonable. One might well expect, for example, that within a family an innately intelligent child might use superior scanning and storage strategies and selectively use certain aspects of his environment that will further develop his intellectual capacity, while a duller sibling in the same family might be selecting less stimulating environmental features in terms of cognitive development. This relationship would produce a positive within-family GE correlation and thus increase apparent genetic influences on the character being measured. Conversely, on many social and personality traits, especially wide phenotypic differences due to underlying genetic expression may be reduced through family pressure toward greater conformity, thus producing a negative GE correlation. While empirical evidence for these two types of GE correlations is scant at this time, if it is found to any degree within families, investigators will be forced to treat such genotype correlated effects as truly genotypic and the residual effects as environmental.

Before returning to the issue of the dependent variables used in most behavioral genetic studies and the implications of this for estimates of gene-environment effects, that special case of genotype-environment study, in which the environment represents age differences, should be mentioned. While a few studies in animal behavior genetics have indicated age by genotype interactions, there is a tendency for certain genetic differences in animal behavior to be remarkably consistent across the wide range of ages. In our own work at Oberlin we have analyzed the genetic pattern of mouse locomotor behavior at several age levels using the diallel cross technique. From weaning upwards the results appear nearly identical, and in all major respects the pattern is the same for mice of two weeks of age who have just recently opened their eyes. Further, we have found a remarkable degree of agreement in the general rank order of various inbred strains used in these studies at ages extending all the way down to four days of age, when young pups are just beginning to locomote, usually in small circular paths.

In human research, Wilson and Harpring (1972) have done extensive analysis of the Bayley scales of mental and motor development with MZ-DZ twin pairs. At each age between three and twenty-four months the twins show substantial within-pair correlations for both mental and motor development and displayed a high concordance for the rate of gain across ages. The correlation for MZ twins approached the reliability of the scale used and were usually significantly higher in the correlation for DZ twins. The results suggest that in any normal home environment the rate of development for each pair of infant twins is primarily determined by a genetic blueprint. At the other end of the age scale Jarvik, Blum and Varma (1972) have presented evidence that genetic components of intellectual functioning are reasonably consistent into old age. The pattern of abilities and concordances of nineteen sets of twins 77 to 88 years of age was still quite similar to that found twenty years earlier with this group, with identical twins still remaining more similar to each other than fraternal twins on most of the measures used. Finally, at present there appears to be no clear picture of changes across ages and various genetic estimates of intellectual abilities obtained in various studies using populations ranging from the early teens through middle age.

What might be considered a special case of genotype by age interaction, genetic by time interactions also deserves mention. In this case we are talking about changes over short time spans, usually in the course of repeated testing or measurement on some particular character such as performance in a learning situation, or repeated tests of some other behavior over a short interval. Fuller and Thompson (1960) in their pioneering book in this field were already aware of the subtleties of this issue in spite of extremely limited data available at that time. Thompson (1966) later expanded his ideas on genotype × time interactions in a discussion of the fluidity of traits both in terms of reliability and stability and from a learning and genetic assimulation standpoint. Estimates of heritability or other genetic components do fluctuate from trial to trial or session to session in many of the studies in which this has been examined. The simplest explanation for this would be that the measures themselves vary in terms of either reliability across sessions or trials, or that ceiling or floor effects are distorting the results, but animal research thus far suggests that these factors are not sufficient to account for the changes in the apparent genetic architecture of a behavioral character across trials. In most cases the shifts are more likely attributable to the fact that one is measuring different aspects of behavior early and later on in such testing situations. The novelty of a test situation on the first few exposures undoubtedly adds an extra dimension to any task that disappears upon repeated testing. Assuming that the behavior being measured is factorially complex, it is likely that the behavior on early trials is factorially different from later trials in a number of test situations.

While some work has been done along these lines perhaps a simpler example of this phenomenon can be seen in a genetic analysis of a shuttle box avoidance learning situation reported by Wilcock and Fulker (1973). Early in the learning session inbred lines of rats showed more avoidance responses than hybrids, whereas on later trials hybrids showed a better avoidance performance. What at first glance appeared to be directional dominance toward poorer learning early and better learning later in the learning session, a radical change in the genetic picture across trials, could be explained rather easily in the light of two factor learning theory. In an avoidance situation, animals must first develop a conditioned fear that is then used as a motivator for subsequent avoidance performance.

Typically, a conditioned emotional response in rats is accompanied by freezing behavior that, of course, led to fewer active avoidances in a shuttle box situation. The apparent poorer performance of hybrids early in learning in the Wilcock and Fulker study thus most likely reflected a more rapid acquisition of a conditioned fear response among hybrids. In later trials their superiority in an operant task was also demonstrated, suggesting directional dominance toward better learning performance for both classical and instrumental aspects of this task, a finding consonant with behavior genetic studies done with laboratory animals. This example is also interesting in that it points out a situation where knowledge of the underlying basis of the behavioral character being measured allowed a parsimonious genetic interpretation of what superficially appeared to be a complex process. Without this understanding, the authors would have been forced to conclude that a rather strange reversal in the genetic structure of learning had occurred in this situation. What is the comparable situation when studying human intellectual abilities? For this one must step briefly into the world of the factor analysts who have been trying to untangle the nature of human intellect.

Factors obtained from the correlations among a number of measurements are, as Burt (1940) long ago pointed out, only statistical abstractions and not concrete entities. Taken alone they are only convenient ways of looking at more complex arrays of interrelationships between measures. Depending on the measures being used, the populations studied, and the assumptions of the model, the factor analyst can come up with a variety of schemes for describing behavior. Thus, Guilford (1967), with reasonably bright, young, and relatively homogeneous samples has, with orthogonal factor analysis and considerable ingenuity in test construction, devised and supported a theory of the structure of intellect that results in 120 independent primary abilities. Earlier, his predecessor, Thurstone (1935), in attempting to avoid a single large G factor came up with seven primary mental abilities. Today, most persons in this field tend to agree that around two-dozen primary factors have been more or less established reliably. Furthermore, most persons in this field tend now to lean toward hierarchical models of intelligence, usually derived from oblique factor analyses where correlations between related factors are in turn factored. Several variations of the hierarchical model do exist, most notably those which result in an overall G factor (e.g., Vernon 1950), those that require more complex relationships among the hierarchies (Humphreys 1962), and Cattell's (1963) hierarchy, which results in a crystallized and fluid higher order intelligence factors.

Obviously, intelligence and aptitude tests that are constructed without reference to such factorial schemes tend to vary considerably in their relative makeup with respect to primary intellectual factors being measured. Because these tests differ from each other in factorially complex ways, it might be argued that genetic studies of intelligence using different IQ tests are not really comparable in terms of the dependent variables being measured. Obviously, when some tests are weighted heavily with primary abilities that are strongly influenced by certain environmental factors, and other tests influenced by primaries that may have a high degree of genetic influence, one would expect somewhat different results when any genetic analyses of the raw test scores are made. One would also expect, and we in fact find, that heritability estimates of various primary factors occasionally run higher and occasionally lower than factorially complex composite tests. There is also no particular reason to think that factor scores, because they are factorially pure, will make more sense or be more sensitive to either environmental or hereditary manipulation than composite tests. While this idea has been expressed, and does have a good deal of intuitive appeal, empirical results so far provide very little support for it. This should not be surprising when we remember that factors are derived from phenotypic correlations that are themselves a function of unknown mixes of a genetic and environmental correlation between measures. Perhaps factor scores based on genotypic or pure environmental correlations would be more useful in this respect, although animal research done along these lines has thus far not shown particularly striking differences between genotypic and environmental correlation matrices.

While factors do not represent entities in themselves, supporting evidence from other sources can begin to strengthen considerably the case for the existence of certain entities originally derived from factor analysis. An example is Cattel's distinction between fluid and crystallized intelligence that emerged as correlated higher order factors when a large number of primary abilities were factored. Semantic relations, formal reasoning, verbal comprehension, mechanical knowledge, and associational fluency tend to load more heavily on the crystallized factor, whereas induction, figural relations, and generally nonverbal skills loaded more heavily on the fluid intelligence factor.

When tests or batteries were made up to tap both crystallized and fluid intelligence, it was found that age curve plots differed for the two aspects of intelligence, with crystallized intelligence continuing to increase slightly with age while fluid intelligence appeared to level off in the early twenties and then to drop gradually (Horn and Cattell 1967). Second, a striking difference in variability in the two abilities seem to exist with fluid intelligence showing much greater variation in a population (Cattell 1940). Finally, some neurological evidence suggests that brain injury at any cortical locality impairs fluid intelligence, whereas only specific aspects of crystallized intelligence are influenced depending on the localization of the injury (e.g., Lashley 1963). From a genetic standpoint, fluid intelligence measures appeared to be more uniform across cultures, and Cattell and associates (Cattell 1971) have reported that fluid intelligence exhibits a greater degree of genetic influence than crystallized intelligence. If all of this evidence proves to be reliable, it would appear that maintaining a distinction between these two related but separable forms of intelligence may be more informative than results based on a single G component and more parsimonious than studying the pattern of inheritance among a large number of primary mental abilities.

Among the relationships between crystallized and fluid intelligence just described, the claim that fluid intelligence primarily involves a genetic component while crystallized intelligence reflects education and other cultural factors, is most likely to be questioned. The data on this point are mixed. Cattell's application of his multiple abstract variance analysis to intelligence data suggested that both fluid and crystallized intelligence were highly heritable although the latter was under somewhat less genetic influence and showed a fair degree of correlation of the action of hereditary and environmental influences (Cattell 1971). Considering the sample size used and the tentative nature of these results, it is difficult to argue that substantial nature-nurture differences have been unequivocally shown for crystallized and fluid intelligence. Some support for this position comes from the ingenious approach of Loehlin and Vandenberg (1968), who correlated difference scores of identical and then fraternal twins on tests from Thurston's primary mental abilities battery and then factor analyzed both the MZ difference data, to produce an environmental factor analysis, and the difference matrix obtained from the subtraction of MZ variance-covariance from DZ variance-covariance matrix, to produce a factor analysis representing mostly genetic effects. Their results indicated that the environmental factors tended to be less intercorrelated than the genetic ones and that the correlations of environmental factors tended to reflect factors mainly involving verbal performance, while the genetic matrix produced a rather uniform second order general factor loaded on all tests. The independence of number and space factors from the verbal factors on the environmental matrix fits rather well with the predictions that are to be made from a crystallized and fluid intelligence theory. On the other hand, the genetic factor looks somewhat more like Spearman's G than the fluid factor described by Cattell.

With respect to the question of measurement of intellectual behavior, and particularly Cattell's fluid-crystallized theory, these two components are correlated but distinguishable but it is premature to make any definite statements concerning relative genetic and environmental influences on these two aspects of intelligence. One of the most striking demonstrations that fluid intelligence can indeed be strongly effected by environmental variables comes from the Belmont and Marolla (1973) study of birth order, family size, and intelligence in the Netherlands. The intelligence test used was the Raven Progressive Matrices test, which appears to be a reasonably pure measure of fluid intelligence. The orderly drop in mean IQ on this test as a function of birth order is striking and obviously not due to genotype.

Based on the rather sparse data now available, I would expect future research to show that both aspects of intelligence are heritable to a moderate degree with crystallized intelligence being somewhat less so, showing a far less stable pattern across age and macroenvironments. Crystallized intelligence will probably be shown to be much more susceptible to both genotype-environment interactions and genotype-environment correlations. Furthermore, the environmental components that go into crystallized intelligence will be largely those of cultural, educational, and occupational factors, whereas environmental influences on fluid intelligence will most likely be those involving prenatal and early postnatal factors including disease, injury, nutrition, and possibly some cases of extreme perceptual motor deprivation early in life. I suspect that as we unravel these aspects of intelligence we will find, as Cattell has suggested (1971), that personality factors are hopelessly intertangled with crystallized intelligence, both influencing its general structure and being influenced by it. I would guess that environmental factors influencing crystallized intelligence will be far easier to detect than will those for fluid intelligence. As Jensen (1973) has pointed out, among the early environmental variables usually called upon to explain differences in intelligence scores between different groups, few can account for more than a small amount of difference except in cases of the most extreme environmental deviation. The possibility that some of these variables interact in some multiplicative manner has yet to be examined closely, however.

It has been argued that test makers perpetuate their own image of intelligence through the nature of items they devise. While this is probably less true now than it was thirty years ago, measures of intellectual ability are still extremely restricted in their scope. Almost all of them are paper and pencil oriented with exception of a few limited performance tests for individual testing situations. Relatively little has yet been done to incorporate some of the work of Bruner, Piaget, and others into standard psychometric procedures, to say nothing of the very broad skills of social and intellectual adaptability normally associated with intelligence but rarely measured adequately with paper and pencil.

I have dwelled at length on the problem of measurement of intelligence primarily to point out that the nature and complexity of the dependent variable of interest can often have considerable impact on findings with respect to its genetic contribution. Equally good examples could have been drawn from the areas of personality and psychopathology. Most behavioral constructs are loosely defined by a collection of specific and narrow behavioral measures. Each of these measures is usually an imperfect index of the construct we wish to examine while at the same time reflecting other factors irrelevant to us. Each specific measure of a behavioral construct is likely therefore to be influenced by genetic and environmental factors in a slightly different way.

A number of years ago I argued that persons interested in studying the influence of early environmental variables in animal research must develop representative design research strategies in which genotype, rearing environment, age, and test variables all are adequately sampled and partitioned out in analysis to get a complete picture of the main effects and interactions of each of these components (Henderson 1969). A similar situation exists with respect to human research, not just as the study of human behavior genetics but rather as the study of human behavioral plasticity. My own entrance into this field was not through an interest in behavior genetics per se, but was prompted because it became evident that omitting genetic variables and the consideration of test parameters was preventing me from answering certain questions concerning the effects of early developmental experiences in animals. The genetic variable is not the most important variable in my research, it is just an essential variable.

No investigator of human behavior examining genetic variables can or should exclude environmental influence. In fact, the standard strategy of maximizing treatment variance should be applied to the environment in such studies. In a sense the term behavior genetics is perhaps a poor one and less descriptive of the issues on which investigators in this field focus than other terms might be. The development of tools for behavior genetic analysis has gone on at a remarkable pace. In fact, the most sophisticated extensions of both human genetic designs and strategies for the study of human development are not much more than a decade old. Methodologically we appear to be in rather good shape for carrying out substantive studies on a large scale, which will help answer many questions concerning human individual differences. Furthermore, there is probably a reasonable agreement among investigators as to the most promising strategies when cost is no factor. The choice and measurement of dependent variables is a difficult one as it always is when one wishes to study complex intellectual processes. I would hope that in future studies a wide domain of behaviors will be examined in each subject, including some of the readily obtainable physiological and biochemical measures thought to be related to these behaviors.

It is obvious that the sample size and the number of measures required in research of this type involves a considerable cost. Should such work be done at all? Probably not if an investigator's goal is just to go on obtaining more or better heritability estimates. These estimates may be useful for selective breeding programs but relatively little else. Should research continue in this field, and if so what will the significance and usefulness of the results be? Some persons, after earning deserved eminence in other areas of psychology and occasionally other fields, have been quite willing to answer these questions, usually in the public press or popular social sciences magazines. In psychology everyone is an expert, and when it comes to the psychology of individual or group differences everyone is an expert with a cause.

Unless individuals directly involved with research concerned with the interrelationships of gene action and behavior directly address the value of such research, continued support for these endeavors may cease given the high cost of doing proper research and the steady stream of misplaced and often uninformed criticism of social scientists with a strong environmentalist positions. The understanding of gene-environment relationships can lead to a far better allocation of limited environmental resources to maximize potential for all individuals. While this concept has been discussed by population geneticists for many years (e.g., Haldane 1933; Wright 1960), social scientists unfortunately still remain largely uninformed on these issues.

References

Allen, G., and Pettigrew, K.D. 1973. Heritability of IQ by social class: Evidence inconclusive. *Science* 182: 1042-44.

Baltes, P.B.; Baltes, M.M.; and Reinert, G. 1970. The relationship between time of measurement and age in cognitive development of children: An application of cross-sectional sequences. *Human Development* 13: 258-68.

Belmont, L., and Marolla, F.A. 1973. Birth order, family size, and intelligence. *Science* 182: 1096-1101.

Burt, C. 1972. Inheritance of general intelligence. *American Psychologist* 27: 175-90.

_____. 1940. *The factors of the mind*. London: University of London Press.
Cattell, R.B. 1971. *Abilities: Their structure, growth, and action*. New York: Houghton Mifflin Co.
_____. 1940. A culture free intelligence test. *Journal of Educational Psychology* 31: 161-79.
_____. 1953. Research designs in psychological genetics with special reference to the multiple variance analysis method. *American Journal of Human Genetics* 5: 76-93.
_____. 1960. The multiple abstract variance analysis equations and solutions: For nature-nurture research on continuous variables. *Psychological Review* 67: 353-72.
_____. 1963. Theory of fluid and crystallized intelligence: A critical experiment. *Journal of Educational Psychology* 54: 1-22.
_____. 1973. Unraveling maturational and learning development by the comparative MAVA and structured learning approaches. In *Life-Span Developmental Psychology*, eds. J.R. Nesselroade and H.W. Reese. New York: Academic Press.
Eaves, L.J. 1972a. Computer simulation of sample size and experimental design in human psychogenetics. *Psychological Bulletin* 77: 144-52.
_____. 1972b. The multivariate analysis of certain genotype-environment interactions. *Behavior Genetics* 2: 241-44.
Erlenmeyer-Kimling, L. 1972. Gene-Environment interactions and the variability of behavior. In *Genetics, environment and behavior*, eds. L. Ehrman, G. Omenn, and E. Caspari. New York: Academic Press.
Erlenmeyer-Kimling, L., and Jarvik, L.F. 1963. Genetics and intelligence: A review. *Science* 142: 1477-79.
Erlenmeyer-Kimling, L., and Stern, S.E. 1973. Technical comments. *Science* 182: 1044-45.
Fuller, J.L., and Thompson, W.R. 1960. *Behavior genetics*. New York: John Wiley and Sons.
Gottesman, I. 1963. Genetic aspects of intelligent behavior. In *Handbook of mental deficiency*, ed. N.R. Ellis. New York: McGraw-Hill.
Guilford, J.P. 1967. *The nature of human intelligence*. New York: McGraw-Hill.
Haldane, J.B.S. 1933. *Science and human life*. New York: Harper.
Henderson, N.D. 1969. Prior treatment effects on open field emotionality: The need for representative design. *Annals of the New York Academy of Science* 159, (3): 860-68.

Henderson, N.D. 1970a. Genetic influences on the behavior of mice can be obscured by laboratory rearing. *Journal of Comparative and Physiological Psychology* 72: 505-11.

———.1970b. Brain weight increases resulting from environmental enrichment: A directional dominance in mice. *Science* 169: 776-78.

———. 1973. Brain weight changes resulting from enriched rearing conditions: A diallel analysis. *Developmental Psychobiology* 6: 367-76.

Horn, J.L. and Cattell, R.B. 1967. Age differences in fluid and crystallized intelligence. *Acta Psychologica* 26: 107-9.

Humphreys, L.G. 1962. The organization of human abilities. *American Psychologist* 17: 475-85.

Hunt, J.McV. 1961. *Intelligence and experience*. New York: Ronald Press.

Jarvik, L.F.; Blum, J.E.; and Varma, A.O. 1972. Genetic components and intellectual functioning during senescence: A 20-year study of aging twins. *Behavior Genetics* 2: 159-71.

Jensen, A.R. 1970. IQ's of identical twins reared apart. *Behavior Genetics* 1: 133-48.

———. 1973. *Educability and group differences*. London: Methuen.

Jinks, J.L., and Fulker, D.W. 1970. Comparison of the biometrical, genetical, MAVA, and classical approaches to the analysis of human behavior. *Psychological Bulletin* 73: 311-49.

Lashley, K.S. 1963. *Brain mechanisms and intelligence*. New York: Dover.

Loehlin, J.C. 1965. Some methodological problems in Cattell's Multiple Abstract Variance Analysis. *Psychological Review* 72: 156-61.

Loehlin, J.C., and Vandenberg, S.G. 1968. Genetic and environmental components in the covariation of cognitive abilities: An additive model. In *Progress in human behavior genetics*, ed. S.G. Vandenberg. Baltimore: The Johns Hopkins Press.

Scarr-Salapatek, S. 1971. Race, social class and IQ. *Science* 174: 1285-95.

Schaie, K.W. 1965. A general model for the study of developmental problems. *Psychological Bulletin* 64: 92-107.

Shields, J. 1962. *Monozygotic twins*. Oxford: Oxford University Press.

Thompson, W.R. 1966. Multivariate experiment in behavior genetics. In *Handbook of Multivariate Experimental Psychology*, ed. R.B. Cattell Chicago: Rand McNally.

Thurstone, L.L. 1935. *Vectors of mind*. Chicago: University of Chicago Press.

Vernon, P.E. 1950. *The Structure of human abilities*, London: Methuen.

Wilcock, J., and Fulker, D.W. 1973. Avoidance learning in rats: Genetic evidence for two distinct behavioral processes in the shuttle box. *Journal of Comparative and Physiological Psychology* 82: 247-53.

Wilson, R. S., and Harpring, E. B. 1972. Mental and motor development in infant twins. *Developmental Psychology* 7: 277-87.

Wright, S. 1960. On the appraisal of genetic effects of radiation in man. In *The biological effects of atomic radiations*. Washington: National Academy of Sciences.

Commentary I

L. Erlenmeyer-Kimling
New York State Psychiatric
Institute

As should be obvious by now, the theme of behavior genetics is that the development of behavior is much like that of any other phenotypic character. What is involved is the continuing co-action of genotype and environmental input at many different levels and the feedback relationships among genes, among the various levels of environment and among the genetic and environmental contributions. It is understood that whether falling within the normal range or lying at the pathological extremes, variations in both animal and human behaviors reflect not only the reaction ranges of given genotypes (see Gottesman 1968)—that is, the different responses that a particular genotype makes in varied environmental circumstances—but also the interaction of genetic and environmental factors, such that differential responses are elicited from different genotypes to identical aspects of environment.

Furthermore, we know that gene-environment interactions need not be linear, and, indeed, where therapeutic or educational manipulations are concerned, we frequently hope that departures from linearity may be found. Nonlinear interactions mean that the differences in phenotypic expression (or in the phenotypic values of quantitative characters) for any pair of genotypes may widen, diminish, or even reverse ranks from one environment to another (see Erlenmeyer-Kimling 1972). The work of Ginsburg (1967), Henderson (1968), Vale and Vale (1969), and others, however, has called to attention the fact that seemingly uninterpretable arrays of gene-environment interactions may contain meaningful regularities that provide information about the mechanisms underlying the behavioral response. Thus, conceptually at least, the study of interactions seems to be where behavior genetics is "at."

In principle, most workers have long acknowledged that the question which is directed at finding out ". . . whether a certain behavioral character is determined by genes or by the environment, or even [at] partitioning the variance into genetic and environmental variance, is. . . misleading from the point of view of understanding behavior" (Caspari 1967, p. 275). In practice, however, most research in behavior genetics, and especially human behavior genetics, has concentrated on just such questions. Very little work so far has been concerned directly with, or even has been capable of, plotting the matrices of gene-environment interactions relating to human behaviors.

There are a number of reasons for the paucity of data, of course. Specifically, research on people is not easy to do, studies in human genetics are especially difficult, assessment of genetic influences on human *behavior* is even more problematic, and the analysis of gene-environment interactions thus becomes exceedingly complex. Dr. Henderson has pointed out in this chapter, with reference to intellectual abilities in particular, some of the difficulties of doing research that makes meaningful interpretations of interactions possible. The following comments are focused chiefly on issues that arise in the study of psychopathology.

The kind of situation that we are grappling with in connection with most psychopathological conditions is the following: Individuals are categorized on the basis of a set of behavioral, and sometimes other phenotypic, characters and are labeled as being afflicted with schizophrenia, manic-depressive psychosis, alcoholism, minimal brain damage, etc. (The reliability of classification or diagnosis of these individuals is a matter that I will not go into here, except to comment that it is perhaps not so troublesome as some people suggest.) Based on family risk studies, biochemistry, or other kinds of data, we hypothesize that—at least in a large proportion of cases—a genetic predisposition is a necessary, but not sufficient, basis for the development of the disorder in question. We will wish to know more about the nature of the genetic predisposition, of course, and we may pursue such questions at the levels of biochemistry, cytogenetics, etc. But we also need to identify the environmental components that interact with the genetic predisposition. Our question, then, is fundamentally quite simple: What features of environment interact with the postulated genotype (or, possibly, group of genotypes) to produce the phenotypic class, and do *not* produce the same phenotype when they impinge on other genotypes? That, to me, is a much easier question than the one that asks about the interactions of genic and extragenic factors underlying quantitative differences in the scores on IQ tests or other measures of abilities.

Three types of problems that may interfere with or obscure the analysis of gene-environment interactions have to be considered. They are (1) the possible heterogeneity of genotypes that are associated with a particular phenotypic grouping; (2) the possible co-variation of genetic and environmental factors; and (3) the time in the developmental process at which measures are made.

Heterogeneity

More and more frequently it is being discovered in human genetics that what may at first seem to be a relatively cohesive phenotypic grouping may involve two, or even more, separate genetic entities. In fact, we can probably expect that some degree of genetic heterogeneity is the usual situation, rather than the converse. Dr. Omenn (see Chapter 5) stresses the likely ubiquity of heterogeneity. The different genotypes may relate to different parts of the same physiological pathway, but then again they may have no relationship to each other and may achieve their effects through completely different routes, as appears to be the case with various audiogenic seizure-prone strains of mice, and as is certainly the case with human deafness and with mental retardation. Mental retardation offers a prime example of heterogeneity of both genetic and nongenetic sources, with at least thirty to forty known single-gene deficiencies, several types of chromosomal aberrations, and various environmental insults being clearly associated with subnormal intelligence. Unfortunate polygenic combinations are thought to account for a further substantial proportion of the cases of mental retardation. The point here is that the responses of the different genetic conditions to nutritional and other aspects of the environment are not at all the same; we can learn little about the reaction range of, for example, phenylketonuric genotypes from studying microcephaly or Down's syndrome. Galactose-free or low phenylalanine diets probably do not do much constructively for genotypes other than the specific ones that they were designed to aid.

Schizophrenia may be genetically heterogeneous; classical manic-depressive psychosis appears to separate into at least two genetically distinct conditions—the bipolar and unipolar forms—and may be further divisible. Based on pharmacological data (cf. Omenn 1973), the minimal brain damage syndrome of childhood has been assumed to encompass two or more different genetic backgrounds. Differences in the manifestations of Huntington's chorea may depend on different sets of modifying genes. Criminality, alcoholism, and other types of addiction are areas of psychopathology that are beginning to attract some attention in genetics; these are obviously highly heterogeneous clusterings of phenotypes, some of which may have specific genetic bases and many of which probably do not. It is important to note that there will be considerable difficulty in disentangling such conditions and perhaps considerable danger of overgeneticizing.

Just as there are real differences in the efficacy of treatment methods among individuals who are supposedly afflicted with the same disorder, there may be important differences in the kinds of environmental factors that interact negatively with the various underlying genotypes. The analysis of gene-environment interactions may, therefore, be blurred and complicated by heterogeneity. At the same time, however, attempts to analyze such interactions may sometimes provide the very key that is needed to allow genetic entities to be teased apart. We are obviously engaged in a series of boot-strapping operations.

Covariation of Genetic and Environmental Factors

The second issue that warrants comment has to do with the difficulty of distinguishing between *interactions*, in which both the genetic and the environmental variables presumably bear a causal relationship to the development of the behavioral expression under study, and the covariance of genotypes and certain aspects of environment—such as social class, or disturbed family life and distorted intrafamilial communication patterns as covariates of the genetic transmission of a predisposition to schizophrenia. We find ourselves here with the chicken-egg dilemma. Sometimes it is possible to distinguish between the two kinds of relationships by means of four-way designs in which the implicated genotype can be studied both in the presence and absence of the implicated environmental variables—and vice versa. The cross-fostering designs carried out in animal work and suggested by Rosenthal (1970) and others for studying schizophrenia represent this type of attempt to separate the effects of gene-environment interactions and gene-environment covariance. It is often impossible, though, to locate adequate samples to fill all of the cells in such a design.

The links between the genetic and environmental characteristics are frequently much too badly intertwined to allow a clear distinction between interaction and covariance. An example can be found in the hypothesis proposed by Kohn (1973) to account for the repeated observation that schizophrenia occurs at a higher rate in the lowest social classes than in the middle and upper levels of society. Some workers have considered low social status per se to be significant in the development of the disorder, while others have attempted to account for the relationship of class to schizophrenia strictly in terms of covariance by suggesting that affected persons and genetically vulnerable individuals experience downward social mobility and thus wind up in the lower rungs of the social ladder regardless of where they may have started. Kohn points out that both genetic vulnerability and excessive stress probably do occur at disproportionately high rates in the lower levels of society but also that the effects of lower social-class position may tend to impair an individual's ability to cope successfully with stress. It is in the face of such class-correlated impairment, Kohn suggests, that genetically vulnerable persons may be at exceptionally high risk for developing schizophrenia when they encounter severe stress. The model, therefore, is one in which the genetic and environmental variables are thought to be intermeshed in both covariance and interactional relationships. Most of our puzzles in human behavior genetics are likely to require this type of multilayered model.

Time

A third problem area in the analysis of gene-environment interactions is concerned with the cross-sectional slice of biological time that is being investigated. Considerable attention has been given to the critical periods at which events must occur if a particular response—either behavioral or physiological—is to develop, as well as to the sensitive periods at which the organism is maximally vulnerable to specific types of treatments. We know of a great many behaviors for which different genotypes show different sensitive periods; outstanding examples are the strain differences with respect to the timing of sensitive periods found in audiogenic seizure research. Fuller and Collins (1970) have observed that there are even genotypic (i.e., strain) differences in the diurnal rhythm of susceptibility to seizures. One point not often mentioned is that sensitive periods need not be confined to a single interval of time in the life span. Many disease susceptibilities, for example, appear to show periods of heightened vulnerability occurring at several different times over the life span.

There is a particular time-related problem that arises in connection with studying conditions, such as schizophrenia and other mental disorders, that become manifest quite late in development. This creates a dilemma if some of the most relevant variables that interact with genetic influences are thought to be those that originate in the individual's earliest life experiences such as parent-child relationships, early peer relationships, and so on. We are then usually dependent on retrospective data from many years before the appearance of the illness in order to learn how the experiences of premorbid and normally-developing people differed, or to learn how and when reactions to perhaps quite commonplace stresses and events began to differentiate the individuals who were later to become mentally disturbed.

For this reason, a number of investigators, particularly in schizophrenia research, have turned to studies of populations in infancy, childhood, adolescence, or early adulthood that are considered, by some criterion, to be at high risk for the disorder. The subject populations are usually selected on genetic grounds (specifically, children of affected parents) because the empirical risks have already been established and have been found to be relatively high. In some studies, selection of a risk group is based on other criteria, such as behavioral traits that are hypothesized to be premorbid characteristics of the disorder in question. However the subject population is chosen, the advantage of the high-risk approach is that the developmental histories of the subjects can be followed prospectively, so that antecedents and consequences of the illness itself can be separated as they never can be in retrospective research. Longitudinal prospective studies are a way to beat the time problem: They open up possibilities of disentangling gene-environment interactions that lie at the roots of the mental disorders of adulthood.

Dr. Henderson called attention to the many complicating and disheartening aspects of undertaking research in human behavior genetics. Longitudinal work, in particular, is notoriously difficult; it is slow to bear fruit, it requires long-term commitments of an investigator's research career, and, it depends on the availibility of major—and, most important, continuing—commitments to support. Thus, whether longitudinal, prospective research on disorders such as schizophrenia will actually be carried out (that is, be started *and* continued) in sufficient magnitude to yield the answers we seek remains questionable. It is encouraging, though, that more than a dozen research groups in the past few years have at least *begun* longitudinal programs on subjects at high risk for schizophrenia.

It is frequently suggested that it is incumbent on the environmentalists to develop better environmental models. True, but it looks as though, if behavior geneticists want to learn more about gene-environment interactions, we will probably have to set up the models for assessing environmental variables ourselves.

References

Caspari, E. 1967. Introduction to Part 1 and remarks on evolutionary aspects of behavior. In *Behavior-genetic analysis*, ed. J. Hirsch. New York: McGraw Hill.

Erlenmeyer-Kimling, L. 1972. Gene-environment interactions and the variability of behavior. In *Genetics, environment, and behavior: implication for educational policy*, eds. L. Ehrman, G. Omenn and E. Caspari. New York: Academic Press.

Fuller, J.L., and Collins, R.L. 1970. Genetics of audiogenic seizures in mice: A parable for psychiatrists. *Seminars in Psychiatry* 2: 75-88.

Ginsburg, B.E. 1967. Genetic parameters in behavioral research. In *Behavior-genetic analysis*, ed. J. Hirsch. New York: McGraw Hill.

Gottesman, I.I. 1968. Biogenetics of race and class. In *Social class, race and psychological development*, eds. M. Deutsch, I. Katz and A.R. Jensen. New York: Holt.

Henderson, N.D. 1968. The confounding effects of genetic variables in early experience research: Can we ignore them? *Developmental Psychobiology* 1: 146-52.

Kohn, M.L. 1973. Social class and schizophrenia: a critical review and a reformulation. *Schizophrenia Bulletin* 7: 60-79.

Omenn, G.S. 1973. Genetic approaches to the syndrome of minimal brain dysfunction. In *Minimal brain dysfunction*, eds. F.F. de la Cruz, B.H.

Fox, and R.H. Roberts. *Annals of the New York Academy of Sciences* 205: 212-22.

Rosenthal, D. 1970. *Genetic theory and abnormal behavior*. New York: McGraw Hill.

Vale, J.R., and Vale, C.A. 1969. Individual differences and general laws in psychology: A reconciliation. *American Psychologist* 24: 1093-1108.

Commentary II

Merrill F. Elias
Syracuse University

If this commentary fails to address itself to all of the interesting points made by Dr. Norman Henderson in "Gene-environment Interaction in Human Behavioral Development," it is because I have been asked to frame my comments in the context of an inquiry into the usefulness of infrahuman animal behavior genetics for the study of behavioral changes over the life span. In view of the rapidly advancing state of the art with regard to the understanding of genetic mechanisms of human behavioral development and improved biometric tools available to the behavior geneticist, one may indeed be concerned about the appropriate role of animal behavior genetics in the study of developmental behavioral processes.

Gene-environment Interactions

Continued demonstrations of gene-environment interactions provide, at best, further evidence for the well-established fact that there are genetic and environmental components of behavior. Thus, many animal behavior geneticists have progressed from studies of inbred strains, characteristic of the early period of animal behavior genetics, to genetic manipulations that enable the detection of single genes or major genes which affect behavior and intervening physiological processes. In some cases it has been possible to relate one or a few genes to a behavioral phenomenon and it has also been possible to trace the intervening physiological mechanisms linking genes and behavior. Wilcock (1969) and Thiessen (1971), among others, have provided critical reviews of the early literature. More recent examples of this approach are provided by research at The Jackson Laboratory (Castelano, Eleftheriou, and Oliverio 1974; Eleftheriou, Bailey, and Denenberg 1974; Messeri, Eleftheriou, and Oliverio 1974; Oliverio, Eleftheriou, and Bailey 1973; Sprott 1972, 1974). These experimenters made excellent use of the recombinant-inbred strains and congenic lines developed by Bailey (1971). These papers are well worth reading in that they demonstrate ways in which the use of the recombinant-inbred strains and the congenic lines permits the experimenter to progress from identification of strain differences to the computation of linkage relationships for each new behavioral or physiological trait that is examined.

Value of Single Gene Research

Research that deals with one, or at best a few genes, has been referred to as single gene research. This approach, characterized by the previously cited literature, has been criticized by behavior geneticists and developmental psychologists. It has been argued that most behaviors of importance to healthy human beings are influenced by many interacting genes and that behaviors which are influenced by only a few genes are qualitatively different and usually far more complex than behaviors typically measured in animal experiments. These arguments question the usefulness of animal models for behavioral genetic research and for behavioral research in general. The value of single gene research has been debated elsewhere (Thiessen 1971; Wilcock 1969) and the legitimacy of animals for behavioral research has been defended in previous papers (Harlow, Gluck and Suomi 1972). However, one of the strongest arguments for the use of animals in behavioral research has not been emphasized sufficiently, that is, the use of animal models for study of the influence of genetic mechanisms on behavioral changes over the life span.

Life-span Behavior Genetics

The life-span approach to developmental psychology assumes that understanding of behavior at any stage of development, for example, old age or senility, may be facilitated by understanding of its historical antecedents. Thus, an investigator derives the best understanding of behavioral changes with age if he follows a behavioral process or processes throughout the life span or at least through a major life segment (Baltes 1968).

As Dr. Henderson suggested, longitudinal data collection in a behavioral genetic context is an expensive, ambitious, frustrating, time-consuming, and often impossible objective to achieve with human subjects. This is particularly true when one adds to the many problems of life-span data collection the dimensions of human and animal behavior genetic analyses that involve extremely large and specialized samples and, as Dr. Henderson advocates, many behavioral measures. Enormous samples are needed just to meet the minimum criterion of three time points and three cohort groups in Dr. Schaie's three-dimensional model (Schaie 1968) for the study of developmental change. Sample size decreases with age as a function of subject attrition (Riegel, Riegel, and Meyer 1968), and in a multifactoral system, subject attrition from factors such as death and illness may itself be influenced by genetic factors. Furthermore, the kind of interface with the public necessary to maintain a human life-span developmental behavior genetics program year after year may challenge the skills of the best public relations man. Assuming that it could be financed, it is unlikely that such an ambitious project could be completed.

Animal Models as Alternatives to Human Research

In spite of the problems of animal husbandry so vividly expressed in operational terms by Dr. Henderson, small animals provide a convenient alternative to human subjects for life-span behavior-genetic studies. In addition to their many assets in terms of experimental control that were noted by Dr. Henderson, small animals have unique advantages for life-span studies. They cannot drop out of the experiment voluntarily. Genetic factors may be manipulated. Most important, their life span is appreciably shorter than that of the experimenter.

These virtues of animal research, and those extolled and illustrated by Dr. Henderson, are given very little weight by investigators who maintain that expediency, relative convenience, and increased environmental and genetic control do not constitute legitimate rationale for the use of animals in life-span behavioral-genetic experiments if the questions answered have no relevance to behavioral changes associated with human development. Undoubtedly, a wide range of research programs with animals can be defended as relevant to human development, particularly by the skilled grantsman. Clearly, a priori decisions with regard to what is and what is not relevant are dangerous. Nevertheless, some areas of investigation are more highly relevant than others, and it may be appropriate for some areas of animal behavior-genetics research to precede studies at the human level.

Genetic Selection, Disease Processes, and Behavior

An area of research that is ripe for the expertise of the "life-span developmental behavior geneticist" is that of disease processes and behavior. Serious questions are now being raised as to the extent to which intellectual decline, except for a sudden drop before death (Riegel and Riegel 1972), is characteristic of healthy aging persons (Baltes and Schaie 1974; Blum, Jarvik, and Clark 1970). Hypertension and cardiovascular disease provide excellent examples of disease processes that, if untreated, can have deleterious effects on cognitive processes and mental health in middle-aged and elderly subjects (Simonson 1965; Spieth 1965; Wilkie and Eisdorfer 1971). Wilkie and Eisdorfer (1971) concluded their report on longitudinal research with regard to blood pressure and intelligence as follows: ". . . the presence of large numbers of aged with cardiovascular illness suggests that the basis for the cognitive decline associated with aging after maturity should be considered secondary to some pathological process and not merely as a 'normal' aging process" (p. 962). Human studies, while extremely important from a descriptive point of view, do not permit the exact control of life history and genotype that is necessary in order to determine the mechanisms involved in the relationship between behavior impairment and untreated hypertension or heart disease. Studies of the relationship between behavior and blood pressure or behavior and heart disease are unavoidably confounded with variations in medication, diet, situational stress, and other life history variables.

One solution to the problem of studying disease processes and behavior is to utilize the genetic selection experiment to develop animal models for disease processes that affect behavior. My definition of an animal model may be clarified in the context of an illustration of a genetic selection program designed to produce blood pressure extremes in mice. Two-way selection for blood pressure extremes and derivation of a randomly mated central group (or groups),from a common foundation stock is the first step. Examination of segregating populations from crosses of selected lines is necessary to determine whether the observed relationships between blood pressure and behavior are based merely on a fortuitous association of genes for behavior and genes for hypertension that responded in a correlated manner during selection. The next step is to identify physiological correlates of behavioral changes and sustained blood pressure extremes that may emerge with the passage of time and constitute the link(s) between genes, blood pressure, and behavior. Of course, investigation of the specific nature of the biochemical lesion causing the hypertension is of importance. The nature of the lesion affects the physiological and behavioral consequences of the disease. Identification of the lesion also provides evidence for the usefulness of the model for certain classifications of human hypertension. Differences in the biochemical link between gene and disease for man and animal model do not necessarily render the animal model useless as many physiological links between the disease and specific behaviors may be similar.

In the example provided, genetic selection is used to study the relationship between disease processes and behavior. In this instance selection is used as a tool in the development of a model for hypertension (phenotype model) rather than as a model for understanding the genetic mechanisms of the disease (genotype model), although the latter objective is not incompatible with the former.

Hypertension is just one example of a disease process that is deserving of investigation at the animal behavior-genetic developmental level. Furthermore, selection need not be the method for developing a model. A variety of genetic materials are available to the developmental psychologist. The mouse mutants, particularly the neurological mutants

(Elias and Eleftheriou 1972; Sidman, Green, and Appel 1965) have not been fully exploited by developmental psychologists. In a recent paper Russell and Sprott (1973) describe many of the neurological mutant mice available for research and discuss their potential use in aging studies, and as noted previously, recent papers by Eleftheriou and his associates emphasize the potential contribution of Bailey's recombinant-inbred strains and congenic lines for the development of genetic models and the establishment of linkage.

There are many other examples of genetic methods and materials (Manosevitz, Lindzey, and Thiessen 1969). The important point is that developmental psychologists can make good use of genetic methods and materials to develop phenotype and genotype models that will enable the characterization of relationships between correlated physiological and behavioral changes with the passage of time. They may do so for physiological processes, normal or abnormal, that are genetically influenced and in turn influence life-span and behavioral change. This is a very different strategy than that of merely characterizing environmental and genetic interactions at different points in the life span, although characterization of environment-gene interactions may constitute an important first step in the development of genotype and phenotype models.

In collaboration with associates at Upstate Medical Center and Syracuse University,[a] Dr. James Florini and I have begun the task of developing an animal model for studying the relationships among hypertension, heart hypertrophy, and learning for animals that are exposed to chronic stress at various life segments during the entire life span. For our initial investigations we are using high and low blood pressure mice and random controls developed at the University of Kansas (Schlager 1974). The eight-way cross developed by Roderick and Wimer has been described by Fuller and Geils (1973). The virtue of Schlager's BP I lines is that randomly mated controls from the same foundation stock are available for comparison with high and low blood pressure lines. Absence of this kind of a control group in many experiments prevents evaluation of the success of the selection experiment and permits no evaluation of the relative position of the high and low lines with regard to a control group (see Elias and Schlager 1974). For example, questions have been raised concerning appropriate biochemical controls for the Okamoto hypertensive rat strains (see Eichelman, Dejong, and Williams 1973).

In addition to maintaining and testing of Schlager's high, low, and random control lines we are developing sublines of inbred strains (BP1H/ES and BP1L/ES) from mating pairs provided by Schlager. These

---

[a]Drs. David Streeten and Gunnar Anderson are examining Schlager's lines in the context of their studies of the angiotensin-renin system at Upstate Medical Center and Dr. Penelope Kelly Elias and Ms. Rosemary Sorrentino are working with me at Syracuse University.

strains and stocks may provide a very useful model within the mouse species for studies of a variety of biochemical, physiological, and behavioral phenomena associated with spontaneous hypertension. Moreover, they can provide a useful experimental prototype for hypertension. (For further reviews of animal models of hypertension, see Gibson 1972 and Schlager 1972).

Expanding Concepts of Developmental Research

The increased restrictions on the use of humans in behavioral research may (in the not too distant future) reduce options for human studies to the extent that investigators will be forced to use animal models in increasingly innovative ways. The unfortunate equating of animal research with comparative, experimental, and physiological psychology for purposes of graduate education has contributed significantly to conceptual and language barriers that have prevented the use of animals in ways in which they can contribute in a unique manner to developmental psychology. Animal behavior genetic research in a life-span context, particularly research dealing with disease processes and behavior, can make a contribution to developmental psychology that goes beyond the description of gene-environmental interactions at different ages.

References

Bailey, D.W. 1971. Recombinant inbred strains. *Transplantation* 11: 404-7.

Baltes, P.B. 1968. Longitudinal and cross-sectional sequences in the study of age and generation effects. *Human Development* 11: 145-71.

Baltes, P.B., and Schaie, K.W. 1974. Aging and IQ: The myth of the twilight years. *Psychology Today* 7: 35-9.

Blum, J.E..; Jarvik, L.F.; and Clark, E.T. 1970. Rate of change on selective tests of intelligence: A 20-year longitudinal study of aging. *Journal of Gerontology* 25: 171-76.

Castelano, C.; Eleftheriou, B.E.; and Oliverio, A. 1974. Chlorpromozine and avoidance: A genetic analysis. *Psychopharmacologia*. In press.

Eichelman, B.; Dejong, W.; and Williams, R.B. 1973. Aggressive behavior in hypertensive and normotensive rat strains. *Physiology and Behavior* 10: 301-4.

Eleftheriou, B.E.; Bailey, D.W.; and Denenberg, V.H. 1974. Genetic analysis of fighting behavior in mice. *Physiology and Behavior*. In press.

Elias, M.F., and Eleftheriou, B.E. 1972. Reversal learning and RNA labeling in neurological mutant mice and normal littermates. *Physiology and Behavior* 9: 27-34.

Elias, M.F., and Schlager, G. 1974. Discrimination learning in mice genetically selected for high and low blood pressure: Initial findings and methodological implications. *Physiology and Behavior* 13: 261-7.

Fuller, J., and Geils, H. 1973. Brain growth in mice selected for high and low brain weight. *Developmental Psychobiology* 5: 307-17.

Gibson, D.G. 1972. *Development of the rodent as a model system of aging.* Washington: DHEW Publication No. (NIH): 72-121.

Harlow, H.F.; Gluck, J.P.; and Suomi, S.J. 1972. Generalization of behavioral data between nonhuman and human animals. *American Psychologist* 27: 709-16.

Manosevitz, M.; Lindzey, G.; and Thiessen, D., eds. 1969. *Behavioral Genetics: Method and research.* New York: Appelton-Century-Crofts.

Messeri, P.; Eleftheriou, B.E.; and Oliverio, A. 1974. Dominance behavior: a phylogenetic analysis in the mouse. *Physiology and Behavior.* In press.

Oliverio, A.; Eleftheriou, B.E.; and Bailey, D.W. 1973. A gene influencing active avoidance performance in mice. *Physiology and Behavior* 11: 497-501.

Riegel, K.F., and Riegel, R.M. 1972. Development, drop and death. *Developmental Psychology* 6: 306-19.

Riegel, K.F.; Reigel, R.M.; and Meyer, G. 1968. A study of the dropout rates in longitudinal research on aging and the prediction of death. In *Middle Age and Aging*, ed. B.L. Neugarten, pp. 563-70. Chicago: University of Chicago Press.

Russell, E.S., and Sprott, R.L. 1973. Genetics and the aging nervous system. In *Survey report on the aging nervous system*, ed. G. L. Maletta. Washington, D.C.: U.S. Government Printing Office.

Schaie, K.W. 1968. Age changes and age differences. In *Middle Age and Aging*, ed. B.L. Neugarten, pp. 558-62. Chicago: University of Chicago Press.

Schlager, G. 1972. Spontaneous hypertension in laboratory animals. *Journal of Heredity* 63: 35-38.

_____.1974. Selection for blood pressure levels in mice. *Genetics*. In press.

Sidman, R.L.; Green, M.C.; and Appel, S.H. 1965. *Catalog of the Neurological Mutants of the Mouse.* Cambridge: Harvard University Press.

Simonson, E. 1965. Performance as a function of age and cardiovascular disease. In *Behavior, Aging and the Nervous System*, eds. A. Welford and J. Birren. Springfield: Thomas Publishing Co.

Spieth, W. 1965. Slowness of task performance and cardiovascular diseases. In *Behavior, Aging and the Nervous System*, eds. A. Welford and J. Birren, Springfield: Thomas Publishing Co.

Sprott, R.L. 1972. Passive avoidance conditioning in inbred mice: Effects of shock intensity, age, and genotype. *Journal of Comparative and Physiological Psychology* 80: 327-34.

Sprott, R.L. 1974. Passive-avoidance performance for single-locus inheritance. *Behavioral Biology* 11. In press.

Thiessen, D.D. 1971. Reply to Wilcock on gene action and behavior. *Psychological Bulletin* 75: 103-5.

Wilcock, J. 1969. Gene action and behavior: An evaluation of major gene pleiotropism. *Psychological Bulletin* 72: 1-29.

Wilkie, F., and Eisdorfer, C. 1971. Intelligence and blood pressure in the aged. *Science* 172: 959-62.

3

Empirical Methods in Quantitative Human Behavior Genetics

John C. Loehlin
The University of Texas at Austin

Quantitative behavior genetics involves the fitting of gene-environment models to continuous behavioral phenotypes. That is, we have in some population a continuously varying behavioral trait, such as intelligence or impulsivity or visual acuity; we have some plain or fancy theoretical model of the underlying genetic and environmental processes that might be responsible for the variation we observe in this trait; and we either try to see how well the model fits the data, relative to some other model, or we take the model as given and estimate some parameter, like gene frequency, heritability, or the relation of within- to between-family environmental variance.

In principle, quantitative behavior genetics could include the fitting of a developmental gene-environment model to the variation of a trait over time in an individual. Advances in developmental genetics and developmental psychology may one day permit this, but in the present chapter the genetics is population genetics, not developmental genetics, and the behavioral variation is variation across people, that is, individual differences. Individual differences assessed at any point in time reflect developmental processes prior to that time—gene-environmental models are in a very basic sense inherently developmental—so that a focus on individual differences does not in any sense slight the general theme of this book.

The topic of quantitative behavior genetics is further limited in the present chapter in two ways: First, the emphasis is on empirical methods—procedures for testing gene-environment models in the real world—rather than on the models themselves. Second, these empirical methods are discussed in the context of quantitative *human* behavior genetics. This does not make much difference to the underlying theory, but it obviously makes a great deal of difference to the methods and designs that can be employed in giving that theory empirical reference.

In short, in this chapter we are concerned with the understanding of human behavioral traits by relating them to underlying gene-environment models, and the emphasis will be empirical—on methods that have been or might reasonably be employed to confront particular models with real data.

Suppose you have a trait that interests you, let us say impulsivity, and you know how to measure it in individuals in some population, and you want to subject this trait to a behavior-genetic analysis. As a first step you would like to break down the total phenotypic variation (that is, the variance of the measured trait) into components associated with differences among the genotypes of the individuals who constitute the population, and with differences among the environments to which they have been exposed. You will soon discover that you need to provide an additional category, to allow for effects that are not properly classifiable as either genetic or environmental. Thus, some of the variance in impulsivity in the population may result from the fact that certain genotypes and certain environments tend to occur together, that is, a gene-environment *correlation* exists. For example, children who have parental models who are high or low in impulsivity will typically receive their genes from the individuals who provide these models. Such a positive gene-environment correlation will tend to increase the range of impulsivity in the population over what it would be otherwise, since gene-environment combinations favoring the development of extreme values of the trait will occur more often. The added variance that results from such a correlation of genes and environments cannot properly be said to be either genetic or environmental—it is both.

Likewise, the population variance may be affected by the fact that genes and environments may *interact,* in the statistical sense; that is, that certain gene-environment combinations may influence the trait in idiosyncratic ways. If children of certain genotypes are made more impulsive by the example of impulsive parents, and children of other genotypes are made less so, there will be present a certain amount of variance that cannot be predicted from either the environment or the genes separately, but only from them jointly.

Errors of measurement might also be placed into this neutral category. There is no obvious reason why defects in the ruler by which one is measured should be attributed to either the genetic or environmental influences upon his growth.

It might be mentioned in passing that the existence of this third category of variation is a common source of disagreement in the heritability estimates of dedicated hereditarians and environmentalists. Each group is inclined to count this middle territory in favor of his side. Also, empirical designs often are differentially sensitive to confounding by effects in this category. As a rough rule of thumb, in empirical assessments gene-environment correlation is most likely to wind up confounded with genetic effects, and gene-environment interaction or errors of measurement with environmental effects—but it can be more complicated than this in particular cases.

Once you have made a broad breakdown of variance into the three major categories, you may well wish to make finer differentiations within them. You may wish to separate out the so-called additive component of the genetic variation (the variance that is "heritable" in the narrow sense, and accounts for resemblance across generations) from genetic variation deriving from dominance relations among alleles at particular chromosomal loci, or from the interactions of genes at different loci. Or you might wish to distinguish genetic variation occurring within human racial or socioeconomic groups in your population from genetic variation occurring between such groups.

Likewise, in the case of environmental variation you may wish to distinguish environmental effects that are common to siblings from environmental effects that differ from child to child within a family, or you may wish to separate out more specific categories of environmental variation, such as that associated with particular nutritional defects or child training procedures.

And, of course, you might wish to break down effects in the third category as well—kinds of interaction, sources of error, and so on.

The program just outlined has yet to be carried out in a convincing fashion for any human behavioral trait. General intelligence is the only trait that even comes close, and the empirical results there are discordant enough so that experts can differ over a substantial range in their estimates of such basic values as the proportions of genetic and environmental variation—let alone such niceties as genetic dominance and gene-environment correlation. A substantial part of the problem is simply inadequate sample sizes. For stable quantitative estimates of genetic and environmental parameters *large* samples are needed—decidedly larger samples than most psychologists are used to (Eaves 1972). The majority of the studies to be mentioned in this chapter, including some of my own, are unsatisfactory in this regard, but they are mainly meant as illustrative.

A general empirical strategy for testing gene-environment models in human populations is easily described: Find subsets of individuals in whom genetic resemblance and environmental resemblance are imperfectly correlated, and see to what degree trait resemblance follows the one or the other. The study of identical and fraternal twins, of adopted children, and of full- and half-siblings, provide examples.

Let us begin with the most widely employed and time-honored of these methods: the comparison of the resemblances of identical and fraternal twins. Here are pairs of individuals of the same age, of the same sex (if only like-sex fraternal twins are employed, as is usual), growing up together in the same family, with the same parental models, the same configuration of ages and sexes of siblings, subjected (usually) to essentially the same child-training procedures, and so on. And yet these pairs of indivuials differ markedly in their degree of genetic resemblance: Identical twins are monozygotic, descended from a single fertilization of an ovum by a sperm, and hence are genetically identical. Fraternal twins are dizygotic, descended from the fertilization of two ova by two sperm, and hence are no more similar genetically than ordinary siblings, who share on the average one-half their genes (if their parents are unrelated). If one assumes that the degree of environmental resemblance is the same for identical and fraternal twins, the difference in the extent to which they resemble one another can be taken as an estimate of the within-family genetic variance, or roughly one-half of the total genetic variation of the trait.

Let us consider some of the methodological "ifs" involved here. First, it is assumed that twin pairs can be accurately diagnosed as mono- or dizygotic. To the extent that this assumption is false, there will be some monozygotic twins in the "fraternal" twin group, making them more similar genetically than they should be, or some dizygotic twins in the "identical" twin group, making them less similar genetically than they should be; either of these errors will decrease the observed identical-fraternal difference, and hence decrease the estimate of genetic variation. If the error rate of the diagnostic method is known, it is possible to adjust for this. As a matter of fact, although twin methodologists often spend inordinate amounts of time, effort, and money on diagnosis, this is probably the least of one's worries in a twin study. Very simple questionnaire methods of zygosity diagnosis have been shown in a number of recent studies (Cederlöf et al. 1961; Cohen et al. 1973; Hauge, Harvald, and Fisher 1968; Jablon et al. 1967; Nichols and Bilbro 1966) to agree well over 90 percent of the time with more elaborate and accurate techniques based on blood groups. Other things equal, the 97 or 98 percent accuracy of blood-grouping methods is, of course, preferable—but others things are rarely equal. If the investigator had really sat down and estimated the expected gain produced by the increased accuracy of blood group diagnoses against the cost in sample size and bias imposed by the necessity of obtaining blood samples, the literature might well contain a good many more studies using questionnaire diagnosis.

Another methodological issue that confronts users of twin methods is the extent to which data obtained from twins can legitimately be generalized to the population at large. Clearly, the twin situation is a special one in many ways. Twins are subject to prenatal stresses of various sorts from crowding in the uterus and competition for nourishment; they tend to be born earlier and at lower birth weights than single births, subjecting them to additional hazards. The interpersonal relationships of twins growing up in a family must differ from those of nontwins in many different ways. And yet there are astonishingly few documented behavioral differences between twin and nontwin individuals. There is a well-established average difference in intellectual ability of perhaps five IQ points, in favor of nontwins. Prenatal and birth difficulties may contribute something to this since the heavier of a pair of same sex twins tends to score a little higher on IQ tests then the lighter, particularly if the birth-weight difference is large (Record, McKeown, and Edwards 1970; Scarr 1969; Willerman and Churchill 1967). However, some of the difference may be related to such postnatal factors as amount of exposure to adults or older siblings. In an English study (Record, McKeown, and Edwards 1970), individuals born as twins but reared as single children, due to death of their partner at birth or in early infancy, showed very little retardation. But this was not true for comparable individuals in a recent United States study (Myrianthopoulos et al. 1972), who showed the usual amount of twin retardation; thus, the question remains open.

Other than the ability difference, and a possible slight increase in the frequency of left-handedness among twins, I do not know of any well-established differences between twins and nontwins in behavioral traits. I have spent a good deal of time with data from large samples of twins and nontwins taken from the National Merit Scholarship Qualifying Test (Loehlin and Nichols, in press; Nichols 1965, 1969). Comparisons were possible on an array of questionnaire and rating measures, including the California Psychological Inventory and Holland Vocational Preference Inventory scales, self-ratings on fifty-seven personality and temperament traits, rated importance of different life goals, and assorted variables related to background, attitudes, and interpersonal behavior. For the females *no* difference between twins and nontwins on any of these variables was large enough to yield a biserial correlation of as much as .13 between the twin-nontwin dichotomy and the measure. Among the males four variables achieved this modest status, with correlations between .13 and .18. They were: self-rating as hard working, CPI Self-control and Achievement via conformance scales, and Nichols and Schnell's scale for the first factor of the CPI (which they describe as value orientation or good social adjustment). Thus, there *might* be a difference between twin and nontwin males in favor of a slightly higher degree of socialization in the twin group—the females showed a difference in the same direction, but it was represented by correlations only on the order of .05. However, even this difference is suspect, since the selective factors in the twin and nontwin samples, while similar—both groups were studied by mail questionnaires—were not quite identical; the twins had to pass the extra hurdle of filling out a zygosity questionnaire in order to gain entry into the sample, so they might be more strongly selected for conscientiousness.

But in any case, twins as individuals do not seem to be different enough from nontwins in their general personality and ability characteristics to give one serious pause in generalizing from the one to the other population.

The classical twin method involves another assumption, however, that concerns the twins not as individuals but as members of pairs: This is that the differences in environmental influences on twins are comparable for identical and fraternal twins.

Now there is plenty of evidence that identical twins get treated more alike in many ways than fraternal twins do, and there is also abundant evidence that identical twins turn out to be more alike than fraternal twins on practically any behavioral characteristic measured. What very few studies bother to examine is whether there is any connection between the two. This is perfectly straightforward to do. All one has to do is look within the *identical* twin group to see whether within this group, differences on the environmental characteristic of interest are associated with differences on the trait or traits involved. Let us consider a concrete case—dressing alike. Identical twins are more often dressed alike in childhood than fraternal twins—could this account in part for their greater resemblance in personality? Well, some identical twins are dressed alike and some are not, and unless the ones that are dressed alike are more similar in personality than the ones that are not, there is not too much point in attributing the greater resemblance of identical twins in general to their greater frequency of dressing alike. This issue has been explored by us in the National Merit data. There are *very* low correlations within the identical twin group between personality resemblance and such factors as having been dressed alike, playing together, being in the same school classes, and so on; so low that it is quite implausible that the difference between identicals and fraternals on these variables could be responsible for the greater resemblances of the identicals.

But in my opinion the really crucial environmental resemblance to consider in the case of twins is probably not that represented by more conventional indices of environmental similarity, but rather the interpersonal environment defined by the relations of twins to each other and of others to the twins as a pair. In this case it is not at all clear that the environments of twins are more similar than those of nontwins, or that the environments of identical twins are more similar than those of fraternals, or that the environments of twins living together are more similar than those of twins living apart. Indeed, there are now several studies in the literature suggesting that twins who are living together may obtain less similar scores on personality measures than twins who are living apart from each other (Price—see Shields 1973; Shields 1962; Wilde 1964). This is typically interpreted as a "contrast effect," in which the twins (or those around them) pick out and exaggerate whatever differences there are between the members of the pair. This effect should be minimal in twins who are separated, hence the higher correlation in spite of less environmental similarity in the ordinary sense. On some scales in some studies identical twins seem to show such an effect more strikingly than fraternals, in other cases it is the other way around. The findings (mostly based on small samples) are fairly erratic, but if this sort of thing goes on to any marked degree it poses some serious problems for the interpretation of traditional twin studies—at least until we understand much more clearly than we now do just how such mechanisms work, what domains of traits they affect, and to what extent they may be method-specific (e.g., found in questionnaires but not in actual behavioral measures). But by the same token, studies of twins may prove a useful setting in which to examine such interpersonal mechanisms.

Let us turn now from studies of identical and fraternal twins to that other traditional design for achieving separation between heredity and environment, the study of adopted children. Here we can find genetically unrelated individuals reared together in the same family, or genetically related individuals brought up in separate families. We can examine the resemblance of adopted children to their biological parents and siblings, and to their adoptive parents and siblings, and from the differences between these draw appropriate inferences about the relative influence of the environment and the genes.

The control over genetic variation in an adoption study is approximately equivalent to that of a twin study, but in this case it is the between-family portion of the genetic variation that is involved. Since the variances due to additive and dominance effects of the genes are differently distributed within and between families, the use of both kinds of data theoretically permits the separate estimation of these two genetic components, and this has been attempted in the case of IQ data (Bulmer 1970; Burt and Howard 1956; Fulker 1973; Jencks 1972; Jinks and Fulker 1970). Most of these authors find an appreciable effect of genetic dominance for the genes influencing IQ, a conclusion that is strengthened by the independent finding of slightly lowered IQ in the offspring of marriages between cousins in Japan (Schull and Neel 1965). This outcome is expected if there are dominant genes that influence IQ in an upward direction.

As in the case of twins, the control of environmental factors in adoption studies can present difficulties, and in adoption studies there is likely to be restriction of genetic range as well. As for environment, the families that adopt children are certainly not a random sample of the population; for obvious reasons the parents tend to be somewhat older and to have fewer children than in the case of natural families; they also tend to be selected by adoption agencies for their emotional and economic stability. As for genes, the parents who give up children for adoption are also by no means a random sample of the population. Data recently gathered by Horn (personal communication) suggest that the girls utilizing a church-related home for unwed mothers in the Southwest had quite atypical MMPI profiles, with average T-scores near 60 on such scales as Pa, Pt, Sc, and Ma and near 70 on Pd.

Finally, there is the issue of selective placement. Most adoption agencies try (within limits) to match the child to the adoptive home, by placing the illegitimate offspring of graduate students in the homes of college professors, and the like. Since the agencies are usually juggling other variables as well (physical appearance, religious background); since they often have very limited information about the biological father of the child; and since they are operating at any given time within limited degrees of freedom in terms of available placements, it would be surprising if they typically achieved a very high degree of parent-child correlation on behavioral characteristics. Also, one should be wary of the automatic assumption that selective placement correlations, if they occur, will be positive. While this might be the rule for ability measures, it seems unlikely that an adoption agency placing a child whose biological parents were (say) alcoholics would look for a "nice alcoholic home" to put him in—if anything, one would expect just the opposite to be the case. In any event the effects of selective placement are balanced out in a full adoption design that estimates the between-family genetic variance in two ways: from the correlation between siblings reared in separate families, and from the difference in correlation between unrelated children reared in the same family and siblings reared together. Any selective placement would tend to produce an overestimate of the genetic variance in the one case and an underestimate in the other.

One more thing should be said about adoption studies. They may soon become extinct. Not because of any lack of scientific merit, but because social changes are rapidly decreasing the availability of children for adoption. The increasing use of abortion to terminate undesired pregnancies, plus the greater social acceptance of an unmarried mother raising her own child, may soon virtually eliminate the scientifically most attractive class of adoptions: the placement of an illegitimate child at birth into an adoptive home with no subsequent knowledge of or contact with its biological parents.

The decline of adoption studies may, however, mean better prospects for another scientifically interesting group, half siblings. These are individuals who have one biological parent in common, but not the other. Half siblings share roughly one-fourth of their additive genetic variance, and no dominance variance. Animal breeders find half sib comparisons useful, since they provide a direct route to the estimation of that part of the genetic variation that can be expected to respond to selection, the additive variance. As another attractive feature, a comparison of half siblings related via the mother to half siblings related via the father can provide an estimate of prenatal and postnatal maternal effects on the offspring.

Little use has so far been made of half siblings in human studies, although the potential value of this group is frequently mentioned in the literature. One reason, no doubt, for this dearth of studies is the general untidiness on the environmental side that is likely to characterize half sibs. Homes are rarely broken cleanly at birth, and half siblings will in many cases have had varying amounts of exposure to the biological and stepparent. Furthermore, it would be most hazardous to assume that the successive mates of a given individual constitute independent random selections from the population with regard to either personality or ability traits. Genetic models can handle correlations between spouses, but locating and measuring all the parties involved may present severe practical difficulties.

One example of the use of the half-sib method in a human study is the investigation of Schuckit (1972) with alcoholics. In this particular social group, broken homes—and hence half siblings—are common. Schuckit began with alcoholic probands, and studied ninety-eight of their half siblings. In this half-sibling group there was a reasonably high incidence of alcoholism, about 25 percent, and one could find all combinations of alcoholic and nonalcoholic biological and stepparents and offspring. Very briefly, what Schuckit found was that the alcoholic and nonalcoholic members of this half sibling group were not differentiated by whether they had *lived* with an alcoholic parent figure (about 30 percent had in each case) but *were* differentiated by whether they had a biological parent who was alcoholic (65 percent among the alcoholics, 19 percent among the nonalcoholics), whether they had lived with the biological parent or not.

To be complete it should be added that while these half-sibling data are clear-cut and striking in their suggestion of a genetic basis for alcoholism, Schuckit and his colleagues (Schuckit, Goodwin, and Winokur 1972) have also made a sib-half sib comparison for this same sample that complicates the picture considerably. Full siblings of the alcoholic probands showed no higher incidence of alcoholism than half siblings, contrary to what one would expect on most genetic hypotheses (or most environmental ones, for that matter). The authors conjecture that an overt alcoholic in the family may actually have a deterrent effect on the development of alcoholism in other family members; in any case some more complex hypothesis (or better, more data) would appear called for here.

Finally, let us consider an often-informative but much neglected group in human behavior genetic studies: the ordinary family. It is clear that the genes and environment are to some degree confounded in ordinary families, and hence that twins, adoptees, half sibs, and so forth have special value. But it often is forgotten that genetic and environmental resemblance are unlikely to be *perfectly* correlated in ordinary families, and if they are not, inference becomes possible. For example, two siblings typically share many more early environmental factors than do a parent and a child, and yet the degree of genetic resemblance is about the same in both cases. For another example, fathers resemble their children as much genetically as mothers do, but in many human societies, including our own, they contribute much less to the immediate social environment of the developing child. If one finds a trait where parent-child correlations equal sibling correlations, and father-child correlations equal mother-child correlations, this is pretty strong prima facie evidence that the genes are responsible. It does not necessarily follow that the heritability of the trait is high—that depends on the size of the correlations. But it does suggest that whatever environmental factors influence the trait tend to be uncorrelated with family membership.

Conversely, if mother-child correlations substantially exceed father-child correlations, this is pretty strong evidence of the operation of some sort of environmental factors. A difference in magnitude between parent-child and sibling correlations is less decisive evidence of environmental effects, since the presence of dominance or assortative mating can affect the relative size of these correlations on purely genetic grounds.

Finally, there can be extremely interesting family correlations in the presence of sex linkage. For instance, if you are fortunate enough to be interested in a trait that is affected by a recessive gene located on the X-chromosome, you will expect to find the highest parent-child correlations to be between mothers and their sons and fathers and their daughters; you will expect the mother-daughter correlation to be intermediate; and you will expect the father-son correlation to be zero. It takes a pretty imaginative environmental hypothesis to generate just this pattern of parent-child correlations—and there are additional genetic predictions concerning sibling correlations as well. Parent-child correlations on tests of spatial visualization have been reported by Stafford and several subsequent investigators to tend toward this recessive sex-linked pattern (Bock and Kolakowski 1973; Hartlage 1970; Insel 1971; Stafford 1961). Since tests of other intellectual abilities do not behave in this way, this seems rather impressive evidence for the genetic hypothesis. Again, the correlations are not high: Spatial visualization (as measured) is not a *highly* heritable trait, but apparently ordinary family environmental factors are not much responsible for individual differences in its development.

Given the various interesting inferential possibilities presented by ordinary family correlations, and the fact that ordinary families are in more abundant supply than twins, adopted children, or half siblings, it is astonishing to have to report that outside of a couple of Cattell's studies I know of no published ordinary sibling correlations on a modern personality inventory, and only one parent-child set—Hill's and Hill's (1973) twenty-eight pairs on the MMPI. Abilities fare a little better—at least IQ does—and now with the interest in spatial visualization we may expect differential abilities to come in for more family studies. It *is* true that correlating persons of different ages (like parents and children) on psychological traits presents many tricky methodological problems. But it is perfectly possible in principle to get two- or even three-generational data from adults, and there seem always to be brave souls who are willing at least to give the same *names* to traits measured in adults and children. Finally, anyone who has access to a public school system or a standardized testing organization and who is willing to wait for two or three years should be able to generate a large number of sibling pairs tested at identical ages. If he is willing to wait for five years, he can compare adjacent to nonadjacent sibling pairs—a strong test for the presence of environmental effects. Paul Nichols (Nichols and Broman 1973) has reported data on this point from sibling pairs in the NIH Collaborative Study, and found nonadjacent sibling pairs to have appreciably lower IQ correlations (by about .10). This was with the Stanford-Binet at age four, when heritabilities may well be lower than later. Nichols will presumably in due course have seven-year WISC scores on this group. But it would be nice to have later school-age figures as well, and data on a much wider range of personality and ability traits.

To summarize, in this chapter we have reviewed the general strategy of apportioning the variation of human behavioral traits into various genetic and environmental and other categories, and considered some of the main empirical methods of doing this with human subjects: twin studies, adoption studies, half siblings, ordinary family correlations. There are many interesting methods that have not been discussed: chromosomal studies, studies of gene linkage, co-twin control studies, studies of intermatings between racially disparate groups, studies of separated monozygotic twins, studies of the offspring of incestuous matings, etc.

Let me conclude with a general methodological dictum, and a brief mention of one particularly intriguing research strategy. The methodological dictum is that whatever the merits of a study of twins or a study of adoptees (say) a study of twins *and* adoptees is better. Each has methodological weaknesses, as we have seen, but they are not the *same* weaknesses, so that taken together the two kinds of studies are much stronger than either alone. Cattell definitely has the right idea with the multiple groups of his MAVA approach (Cattell, Blewett, and Beloff 1955; Cattell, Stice, and Kristy 1957), whatever minor quarrels one might have with the details of its implementation (Loehlin 1965). Jinks and Fulker (1970) provide many of the necessary statistical refinements—now all that is needed is the data.

Finally, the particularly intriguing strategy is a multivariate one. It turns out that all of the described methods for apportioning the variance of a single trait into various genetic and environmental components can also be applied to the covariances (or correlations) between traits. Now it may often be interesting to know what role the genes and the environment play in accounting for variation in (say) a test of verbal reasoning or a complex spatial relations test. But it may be even more interesting to get a similar breakdown of the variance those two tests have in common: Is the covariance *higher* in genetic loading than the separate tests, as a proponent of an inherited Spearman "g" might expect, or *lower,* as someone might predict who sees broad environmental factors such as education and social class as bringing about correlations among originally independent abilities?

Naturally, there is no reason to stop with the correlations between just *two* measures, and soon one is off gaily factor-analyzing matrices of genetic and environmental covariances (Loehlin and Vandenberg 1968; Meredith 1968). A happy picture. But we must always remember that if *single*-variable human behavior genetic analyses call for large samples, multivariate behavior genetic analyses call for very, very large samples —not exceptionally large in terms of research in some other sciences, though. And large-scale data-gathering and data-processing are certainly technically feasible nowadays. So perhaps the time has come to give a serious trial to methods such as these.

References

Bock, R.D., and Kolakowski, D. 1973. Futher evidence of sex-linked major-gene influence on human spatial visualizing ability. *American Journal of Human Genetics* 25: 1-14.

Bulmer, M.G. 1970. *The biology of twinning in man.* London: Oxford University Press.

Burt, C., and Howard, M. 1956. The multifactorial theory of inheritance and its application to intelligence. *British Journal of Statistical Psychology* 9: 95-131.

Cattell, R.B.; Blewett, D.B.; and Beloff, J.R. 1955. The inheritance of personality. *American Journal of Human Genetics* 7: 122-46.

Cattell, R.B.; Stice, G.F.; and Kristy, N.F. 1957. A first approximation to nature-nurture ratios for eleven primary personality factors in objective tests. *Journal of Abnormal and Social Psychology* 54: 143-59.

Cederlöf, R.; Friberg, L.; Jonsson, E.; and Kaij, L. 1961. Studies on similarity diagnosis in twins with the aid of mailed questionnaires. *Acta Genetica et Statistica Medica* 11: 338-62.

Cohen, D.; Dibble, E.; Grawe, J.; and Pollin, W. 1973. Separating identical from fraternal twins. *Archives of General Psychiatry* 29: 465-69.

Eaves, L.J. 1972. Computer simulation of sample size and experimental design in human psychogenetics. *Psychological Bulletin* 77: 144-52.

Fulker, D.W. 1973. A biometrical genetic approach to intelligence and schizophrenia. *Social Biology* 20: 266-75.

Hartlage, L.C. 1970. Sex-linked inheritance of spatial ability. *Perceptual and Motor Skills* 31: 610.

Hauge, M. et al. 1968. The Danish twin register. *Acta Geneticae Medicae et Gemelloglogaie* 17: 315-32.

Hill, M.S. and Hill, R.N. 1973. Hereditary influence on the normal personality using the MMPI. I. Age-corrected parent-offspring resemblances. *Behavior Genetics* 3: 133-44.

Insel, P. 1971. Family similarities in personality, intelligence and social attitudes. Ph.D. dissertation, University of London.

Jablon, S.; Neel, J.V.; Gershowitz, H.; and Atkinson, G.F. 1967. The NAS-NRC twin panel. *American Journal of Human Genetics* 19: 133-61.

Jencks, C. 1972. *Inequality: A reassessment of the effect of family and schooling in America.* New York: Basic Books.

Jinks, J.L., and Fulker, D.W. 1970. Comparison of the biometrical genetical, MAVA, and classical approaches to the analysis of human behavior. *Psychological Bulletin* 73: 311-49.

Loehlin, J.C. 1965. Some methodological problems in Cattell's Multiple Abstract Variance Analysis. *Psychological Review* 72: 156-61.

Loehlin, J.C., and Nichols, R.C. In press. *Heritability of personality, ability and interests.* Austin: University of Texas Press.

Loehlin, J.C., and Vandenberg, S.G. 1968. Genetic and environmental components in the covariation of cognitive abilities: An additive model. In *Progress in human behavior genetics,* ed. S.G. Vandenberg. Baltimore: Johns Hopkins Press.

Meredith, W. 1968. Factor analysis and the use of inbred strains. In *Progress in human behavior genetics,* ed. S.G. Vandenberg. Baltimore: Johns Hopkins Press.

Myrianthopoulos, N.C.; Nichols, P.L.; Broman, S.H.; and Anderson, V.E. 1972. Intellectual development of a prospectively studied population of twins and comparison with singletons. *Human Genetics: Proceedings of the fourth International Congress of Human Genetics.* Amsterdam: Excerpta Medica.

Nichols, P.L., and Broman, S.H. 1973. Familial factors associated with IQ at four years. Paper presented at Behavior Genetics Association Meetings, April 1973, Chapel Hill, N.C.

Nichols, R.C. 1965. The National Merit Twin Study. In *Methods and goals in human behavior genetics,* ed. S.G. Vandenberg. New York: Academic Press.

_____. 1969. The resemblance of twins in personality and interests. In *Behavioral genetics: Method and research,* eds, M. Manosevitz, G. Lindzey, and D.D. Thiessen. New York: Appleton-Century-Crofts.

Nichols, R.C., and Bilbro, W.C., Jr. 1966. The diagnosis of twin zygosity. *Acta Genetica et Statistica Medica* 196: 265-75.

Record, R.G.; McKeown, T.; and Edwards, J.H. 1970. An investigation of the difference in measured intelligence between twins and single births. *Annals of Human Genetics* 34: 11-20.

Scarr, S. 1969. Effects of birth weight on later intelligence. *Social Biology* 16: 249-56.

Schuckit, M.A. 1972. Family history and half-sibling research in alcoholism. *Annals of N.Y. Academy of Sciences* 197: 121-25.

Schuckit, M.A.; Goodwin, D.A.; and Winokur, G. 1972. A study of alcoholism in half siblings. *American Journal of Psychiatry* 128: 1132-36.

Schull, W.J., and Neel, J.V. 1965. *The effects of inbreeding on Japanese children.* New York: Harper & Row.

Shields, J. 1962. *Monozygotic twins.* London: Oxford University Press.

_____. 1973. Heredity and psychological abnormality. In *Handbook of abnormal psychology,* ed. H.J. Eysenck. London: Pitman Medical.

Stafford, R.E. 1961. Sex differences in spatial visualization as evidence of sex-linked inheritance. *Perceptual and Motor Skills* 13: 428.

Wilde, G.J.S. 1964. Inheritance of personality traits. *Acta Psychologica* 22: 37-51.

Willerman, L., and Churchill, J.A. 1967. Intelligence and birth weight in identical twins. *Child Development* 38: 623-29.

Commentary

Steven G. Vandenberg
University of Colorado

Dr. John Loehlin has explored "Empirical Methods in Quantitative Human Behavior Genetics" so well that amplification of some of the points he made seems the only adequate response. I agree with him when he states that tracing the behavioral consequences of specific genetic abnormalities (or of unusual environmental conditions) is the preferred strategy. Unfortunately, it is very difficult for a single investigator to collect enough cases to perform an adequate analysis. Perhaps we should publish descriptions of individual cases in sufficient detail so that they can be integrated and analyzed together at a later time.

An interesting example of such an analysis is given in a paper by Moor (1967), who summarized the findings about the intelligence of various types of aneuploidies reported in the literature. She found a regular decrease in mean IQ for each excess X-chromosome and a smaller decrease for each excess Y as shown in Figure 3-1. Such an analysis would be still more informative if information on the IQ of both parents and of normal siblings is also obtained. This would permit the use of a linear regression model to obtain coefficients that would show the relative importance of the excess X's and Y's as well as the parental contributions to the proband's intelligence. Perhaps one could also determine whether the decrease in IQ is the same at all expected levels of intelligence, or one proportional to the value expected had the child not been aneuploid. It is possible that such information on patients and their parents already is available in some of the records at various genetic counseling services.

Striking examples of what can be done with data collected for other purposes are provided by the recent analyses of the effect of birth order on intelligence by Record, McKeown, and Edwards (1969) of "eleven plus" test scores in Birmingham, England; by L. Belmont and Marolla (1973) of Raven Progressive Matrices scores of military recruits in Holland; by Breland (1972) of National Merit Scholarship data in the United States and by Vallot (1973) of children tested (with a intelligence test constructed for that purpose) by the French National Demographic Institute. A reanalysis of these four studies may permit a comparison of the percentage of the total variance due to birth order in England, Holland, the United States, and France, which could perhaps suggest looking for social correlates of differences in the magnitude of the effect *if this varies*, or which would suggest either a social factor that is constant across these four countries or a biological factor that would need to be explained, if the effect is of the same

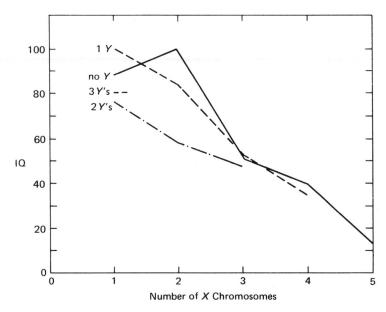

Figure 3-1. Mean IQ of Persons with Various Numbers of X and Y Chromosomes (Moor 1967)

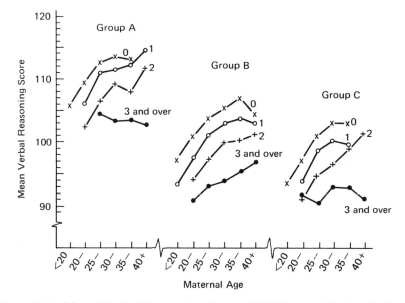

Figure 3-2. Mean Verbal Reasoning Score of Children with 0, 1, 2, 3, or More Siblings, Born to Mothers of Varying Age for Three Socio-Economic Groups (Record et al., 1969)

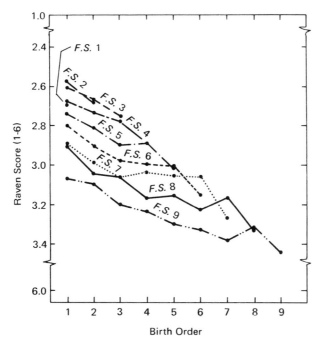

Figure 3-3. Mean Raven Class Score by Birth Order within Family Size (FS) Across Dutch Recruits (N = 386,114) [Belmont & Marolla 1973]

magnitude. In this respect it is interesting to observe that the birth order effect, even though it is small, is independent of family, age of the mother, or socioeconomic status, as can be seen from Figures 3-2 through 3-5 which summarize the results of the four studies. There may even be enough twins in some of these survey studies to permit study of the birth order effect just in twins, to compare the size of this effect relative to that of the hereditary and environmental factors.

Returning to Dr. Loehlin's discussion for another stimulus, I would like to elaborate on his concern for the need of large sample size. Further analyses of already existing data unfortunately limits one to the variables chosen by the original investigators, often for a very different and sometimes a more applied purpose. To obtain new data on samples of adequate size it may be necessary for a number of investigators to cooperate using the same procedures. This raises questions about divided authority and responsibility, as well as multiple authorship, but these problems need not be insurmountable as shown by the cooperation between four persons from the University of Colorado and four from the University of Hawaii.

Dr. Loehlin lists studies of identical and fraternal twins, of adopted

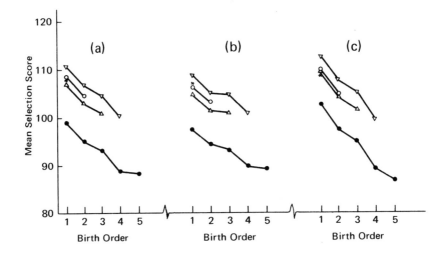

Figure 3-4. A. Mean National Merit Scholarship Selection Score for Students of Different Birth Orders and from Families of Different Sizes
B. Idem, After Correction for Socioeconomic Status (SES)
C. After Correction for SES and Age of Mother (Breland 1972)

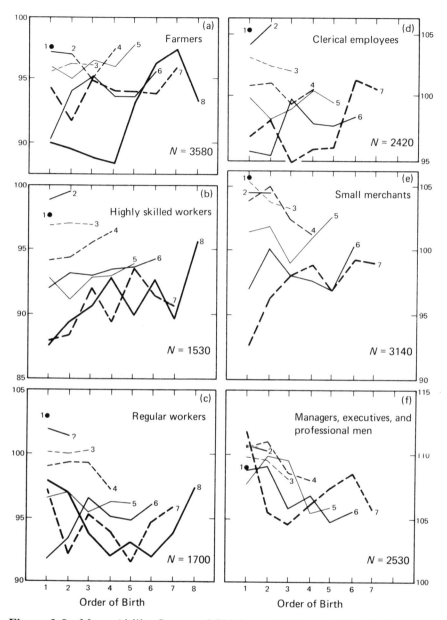

Figure 3-5. Mean Ability Scores of Children of Different Birth Orders and Families of Different Size Whose Fathers are:

    A. farmers
    B. highly skilled workers
    C. regular workers
    D. clerical employees
    E. small merchants
    F. managers, executives, and professional men. (Vallot 1973).

children, and of full and half sibs as examples of the other strategy where normal variance is partitioned into hereditary and environmental portions. I would like to add a few more. Nance (1974) has suggested that we study the children of identical twins. Because such children each have one of the identical twin pair as a parent, they are genetically equal to half sibs, without the disruptive influence of divorce or the systematic age difference between the children of the two different spouses that are found in serial marriages. Occasionally one might even be able to find identical twins married to a pair of identical twins (Taylor 1971). The offspring of such identical couples would be full sibs who may experience between-family environmental differences that can be specified to some extent.

The next point of Dr. Loehlin's to comment on concerns what may be called the "twinness" of different twin pairs. The factors mentioned by Zazzo (1960), von Bracken (1969), Smith (1965), and others who investigated dyadic relations within twin pairs may lend themselves to the construction of a measure of the degree of intimacy and interdependence versus separateness of pairs of twins. If such a variable can be shown to be related to identical as well as fraternal twin differences in ability and/or personality, we may be able to improve heritability estimates based on identical and fraternal twin concordances, as suggested by Loehlin (1973). We could then also attempt to understand how such a variable affects both types of twins. It would also be of special interest to study these factors in boy-girl twin pairs, which are usually excluded from studies. Again, it may be that data of this kind are available from past twin studies.

Dr. Loehlin also mentioned the absence of information about possible differences between twins and single born on ability and personality data. It should not be too difficult to match existing twin data to newly obtained data on single born or to "norms" from test manuals.

My final remarks concern a relatively minor point. In other areas of genetics one hears of scaling or rescaling phenotypes. In human behavior genetics we have not reached that point. Recently I analyzed the responses of twins to a questionnaire about motion sickness. I found a considerable genetic component even though a minority of persons had this problem, while those affected had experienced the problem in different situations such as riding in a car, bus, boat, plane, or on fairground rides.

I looked for all or none concordance or discordance and later considered the magnitude of the discordance. Because the responses to the questions were in categorical form, there is no obvious way to quantify the answers. Let us select an example: To a question about carsickness, the answers were (a) I never get carsick, (b) only on long rides, (c) sometimes on short rides, (d) every time I ride in a car.

In the first analysis any differences in response were scored as discordance; in the second one, the choices were treated as zero, one, two, and three. For many of the items the conclusions regarding the importance of genetic factors were different depending on the type of analysis used. This is somewhat the same point that arises when one does or does not consider the severity of schizophrenia.

The next concern I would like to express is that we are neglecting natural behaviors such as walking, sleeping, eating, and so on. Darwin described the behavior of a man who during his sleep would throw his arm about, frequently hitting his wife. He mentioned that a son and grandson showed similar behaviors. In my family some of the individuals make a peculiar excess movement with one leg when they walk.

The quality of one's voice is very distinct, and family members may be mistaken for one another over the phone. We can study this by comparing "voiceprints," that is, spectral analysis of the frequency of soundwaves in short segments of speech. (For details see Potter, Kopp, and Green 1947). Methods for comparing pairs of voice prints other than by subjective ratings would have to be developed. It will be easy to characterize the duration of each vowel or sibilant and the height of the pitch, by direct measurement on the voice print, but it is less obvious how to capture more holistic attributes.

My final comment concerns environmental factors. In part because of the Equality of Educational Opportunity report by Coleman et al. (1966), an increased interest has developed in the development of measures or indices of environmental factors, either through parental questionnaires or through other means.

Williams (1973) studied the relations of parental attributes to children's and parent's Wechsler intelligence scores by multivariate path analysis and more recently (1974) has compared the environmental indices and the variance they accounted for in four studies: by Wolf (1964) in Illinois; by Marjoribanks (1970) in Eastern Canada; by Mosychuk (1969) in Edmonton, Canada; and by Keeves (1972) in Australia. There is considerable agreement on overall factors and comparable variance accounted for (around 40 percent), but not much overlap in specific details of the questionnaires. In fact, the questionnaires are not empirically tested nor based on any serious theoretical effort. Construction of such indices forms an important problem that warrants more careful work both on the theoretical and the empirical side. Williams found that a social learning model that would suggest three major influences of parents on the cognitive development of their children gave a better fit than the ad hoc dimensions used in these four studies.

The intent of my comments has been a call for better as well as more data, for a search for and exploitation of data collected for other purposes, and for cooperation between two or more investigators at different locations. It is suggested also that we might study walking, voice qualities, sleeping, and other "species-specific" behaviors that show at times interesting idiosyncrasies that are often said to run in families. Finally, mention is made of studies of parental influences on children that could be highly relevant to behavior genetic studies.

References

Belmont, L., and Marolla, F.A. 1973. Birthorder, family size and intelligence. *Science* 182: 1096-1101.

Breland, H.M. 1972. *Birthorder, family configuration and verbal achievement*. Princeton, N.J.: Educational Testing Service, Research Bulletin, 72-47.

Coleman, J.S.; Campbell, E.Q.; Hobson, C.J.; McPartland, J.; Mood, A.M.; Weinfeld, F.D.; and York, R.L. 1966. *Equality of educational opportunity*. Washington, D.C.: U.S. Department of Health, Education and Welfare. Office of Education. Catalog number FS 5.238:38001.

Keeves, J.P. 1972. *Educational environment and student achievement*. Stockholm: Almqvist and Wiksell.

Loehlin, J. 1973. Personal communication.

Marjoribanks, K.M. 1970. Ethnic and environmental influences on levels and profiles of mental abilities. Ph.D. dissertation, University of Toronto.

Moor, L. 1967. Niveau intellectuel et polygonosomie: Confrontation du caryotype et du niveau mental de 374 malades dont le caryotype comporte un exces de chromosomes X ou Y (Intellectual level and polyploidy: A comparison of karyotype and intelligence of 374 patients with extra X or Y chromosomes). *Rev. Neuropsychiat. Infant.* 15: 325-48.

Mosychuk, H. 1969. Differential home environments and mental ability patterns. Ph.D. dissertation, University of Alberta.

Nance, W.E.; Nakata, M.; Paul, T.D.; and Yu, P.L. 1974. The use of twin studies in the analysis of phenotypic traits in man. In *Congenital defects: New directions in research*, eds. D.T. Janerich, R.G. Skalko, and I.H. Porter. New York: Academic Press.

Potter, R.K.; Kopp, G.A.; and Green, H.C. 1947. *Visible Speech*. New York: Van Nostrand.

Record, R.G.; McKeown, T.; and Edwards, J.H. 1969. The relation of measured intelligence to birth order and maternal age. *Annals of Human Genetics* 33: 61-69.

Smith, R.T. 1965. A comparison of socio-environmental factors in monozygotic and dizygotic twins, testing an assumption. In *Methods and goals in behavior genetics,* ed. S.G. Vandenberg, pp. 45-61. New York: Academic Press.

Taylor, C.C. 1971. Marriages of twins to twins. *Acta Geneticae Medicae at Gemellologiae* 20: 96-113.

Vallot, F. 1973. Niveau intellectuel selon le milieu social et scolaire. In *Enquete nationale sur le niveau intellectuel des enfants d'age scolaire,* pp. 25-116. Cahier 64, Institute National d' Etudes Demographiques Paris, Presses Universitaires de France.

von Bracken, H. 1969. Humangenetische Psychologie. In *Humangenetik,* ed. P.E. Becker, vol. 1/2, pp. 409-561. Stuttgart: Thieme.

Williams, T. 1973. Cultural deprivation and intelligence: extensions of the basic model. Ph.D. dissertation, University of Toronto.

———. 1974. Competence dimensions of family environments. Paper presented at the annual meeting of the American Education Research Association, April 1974, Chicago.

Wolf, R.M. 1964. The identification and measurement of environmental process variables related to intelligence. Ph.D. dissertation, University of Chicago.

Zazzo, R. 1960. *Les jumeaux, le couple et la personne* (Twins, the pair and the individual). Paris: Presses Universitaires de France.

4

Populations for the Study of Behavior Traits

G. Ainsworth Harrison
University of Oxford

Apart from a few relatively simply inherited sensory perception traits like color vision and capacity to taste phenyl-thiocarbomide, there is very little information on most of the world's populations concerning behavioral variety that is likely to be of value to geneticists. The types of data collected by social anthropologists and ethnologists, important as it is for their purposes, is not of a form amenable to genetic analysis. They tend to be concerned with behaviors that are likely to have little if any additive genetic variance, but, more importantly, they relate their observations much more to average, if not idealized, behavior patterns rather than to individual variety.

Psychologists, interested in the experimental analysis of behavior, have not totally confined themselves to industrialized societies. Although these have carried the brunt of their concern, a considerable amount of information has been collected, for example, on cognitive abilities in Australian Aborigines and various tribal and urban groups in Africa. A rapidly increasing amount of information is also becoming available for many populations on such ontogenetic characteristics as patterns of motor development in children and the ways in which these are affected by natural environments and cultural practices. It is sometimes possible, when genetically similar and different groups are appropriately distributed with respect to environmental variables, to estimate whether some behavior patterns, especially ones like motor development, have genetic components to their variation (Harrison 1966), but as with the social anthropological data it would appear that little of the psychological information available is of a form suitable for any formal genetic analysis. The main reason for this is, of course, the absence of family data that for developmental characteristics are for all intents and purposes impossible to obtain, especially in preliterate societies. In the case of variation in cognitive abilities one can also doubt whether the tests being used are measuring anything worth analyzing in nonwesternized populations and indeed whether they are even measuring this with any accuracy. This is not an issue of whether tests are "culture fair," which only becomes of concern in dealing with the second order problem of analyzing between-population differences, but rather of whether any test devised outside a culture can accurately detect consistent and meaningful cognitive behavioral differences in it.

The current anthropological situation thus looks bleak and without immediate remedy from the behavioral geneticist's viewpoint. One might be tempted to advocate broad quantitative descriptive studies of patterns of behavioral variety within as many as possible of the world's populations. In general, such global studies have been anthropologically and genetically useful for many physical human characteristics, and are often a prerequisite to more sophisticated and analytical investigations. Further, many human groups are clearly in peril and will probably not be available for study, at least in anything like their present forms, in but a few year's time. However, it would appear that quite apart from the problems involved in any genetic analyses we may not yet be in a position to undertake worthwhile anthropological investigations of interest to behavior geneticists. The biological anthropologists may nevertheless still have something of value to contribute to studies of human population behavioral genetics. This is in the area of defining the population units for investigation and analyzing their microstructure, which is required as much in developed societies as in others.

The choice of a particular population or group of populations for intensive study is primarily dependent upon the specific problems one poses. This is at least as true for behavioral variation as for any other form of variation. Broadly speaking, these problems can be grouped into three main categories: The first concerns questions of the mode of determination of the observed variation, and involves not only interest in the relative roles of environmental and hereditary factors, which, as is well enough known for behavioral characteristics, can interact in extremely complex ways, but also, in principle, the precise modes of inheritance of any genetic element, including the localization of responsible genes.

The second type of problem can be thought of as the questions related to the evolutionary reasons, why variation within and between populations exists and why it takes the form it does. Strictly speaking the evolutionist would only be interested in genetic components to the variation but there may still be considerable value in examining the fitness effects of behavioral variation that are entirely environmental in origin. Anthropologists, for instance, often want to know how culturally determined behavior patterns relate to the occupation of particular ecological situations, and, though little interest as yet has been shown in the effects of behavioral heterogeneity on individual or group survival within any one ecological population, such studies are likely to develop.

The final category of questions concerns the concomitant effects of behavioral variation on other aspects of population structure. Thus, for instance, it is a matter of some concern how behavior affects the geographic movement of people and through such movement clusters Mendelian populations in space and distributes genes. Vertical movement through social systems, either through marriage or through social mobility, is likewise both profoundly affected by behavior patterns and influential in determining the ways populations are clustered.

These groups of questions are of course closely interrelated. The way one approaches the design of heritability studies and other investigations into the nature of any genetic variation should take into account the form of the population structure; while population structure, as it is influenced by movement, depends upon the genetics of the behavior that determines the movement. Evolutionary forces operate within the framework of a population structure and affect the form genetic variety takes. Nevertheless, it is obvious that the choice of populations for behavioral studies must depend upon the problems being addressed.

On the other hand, whatever the specific question of concern, there are some general factors and requirements that need to be taken into account in making a selection. It is a sine qua non for behavioral geneticists that data collection should be directed to family studies and that such studies optimally should embrace as many generations as possible. Nevertheless, few population studies have been orientated this way no doubt because of the comparative ease of examining groups such as children or military recruits as compared with a house to house survey that would be a useful way of obtaining pedigree information. Such family studies demand not only a much greater involvement and commitment on the part of investigators, but also a high level of cooperativeness from potential subjects. On the whole one tends to find this cooperativeness in more rural groups than in urban ones, but confining one's attention to the former necessarily raises the question of their general representativeness. In any event house to house surveys are expensive and raise difficult logistic problems.

Although attention to families is of paramount importance, particularly so that one can obtain information on the basic genetic requirement of groups consisting of pairs of parents and two of their children, it is also of value, especially with respect to evolutionary types of questions, to consider data on unmarried adults, couples who have had no children, as well as single parents or single offsprings. No doubt behavior and fertility are related in the most fickle of ways and one finds dramatic temporal changes in the fertility patterns of different social groups. However, that there are strictly biological causes for infertility is beyond dispute and the extent that these are influenced by behavior patterns is of more than passing interest.

A second consideration in identifying populations for the analysis of behavioral traits concerns decisions about the size of the population required. Here there are a number of conflicting interests. Small demographic groups are likely to correspond most precisely with the basic units in the hierarchy of Mendelian populations—in which, for instance, mating with respect to ancestry may be more or less at random. Many analyses have been based upon this assumption. With small units it is also possible to achieve something approaching total ascertainment of individuals and this avoids problems of sampling that, for variation in behavioral traits, can be especially complex. On the other hand, the situations that prevail in small and isolated groups can be very idiosyncratic and it is dangerous to extrapolate and generalize from them to other situations.

What can be said with certainty is that there is a vital need in any study of the genetics of behavioral variation to know as much as possible about both the present demographic structure and the past demographic history of the population being investigated. Many of the forces of concern to the population geneticist are manifest in demographic characteristics and the results of their action tend to be dependent on time. Of particular concern in the interpretation of observed patterns of variety is whether a population is in equilibrium with the demographic forces operating within it. Distinctive patterns frequently appear before a complete equilibrium state is reached, but nevertheless usually require at least a number of generations to develop. Where possible one would choose to analyze populations that were in an equilibrium state: under such situations one can for instance most easily construct models to test against observation. However, there can be few human groups that today are in equilibrium and the best that one can often hope to establish is how far a situation is from equilibrium and whether there has been sufficient time for any representative pattern to emerge. This requires demographic data of a historical nature with particular emphasis upon changing patterns of population growth and movement.

Another and more specific reason for being concerned with demography relates to the nature of migrants. There are very few, if any, totally closed systems in developed parts of the world and most populations experience an inflow and outflow. In-migrants, however, can usually be identified without difficulty and their effects measured. The main problem is concerned with the nature of out-migrants, who at least for behavioral traits are not likely to be representative samples of the population into which they were born. This can cause severe analytical difficulty. Thus, for instance, to examine the effects of some behavioral attribute on social mobility, the best design is to compare the attribute in parents and their own children in relation to occupation. However, occupation is clearly an attribute of adults and by the time any cross-sectional investigation is undertaken many offspring will have left their parental home; the pattern of leaving is likely to be profoundly affected by occupation. Usually the upwardly mobile are the ones most likely to move. Locating out-migrants is logistically an extremely burdensome problem, but it is evident that they are crucial not only to studies of social mobility but also to more general investigations of the genetics of behavioral variation. Out-migration is particularly bothersome in studying small rural populations, but it has to be taken into account in most other situations as well. The difficulties that can arise as a consequence of out-migration are somewhat overcome by examining samples of larger groups, but for many purposes some form of family identification is still a prerequisite.

Another prerequisite for the fine analysis of behavioral variation and related to the demographic requirement is detailed knowledge of the social system. In Britain, for instance, any study has to take into account the phenomenon of social class and though class structures there are more precisely defined than in most other countries, comparable phenomena exist in other societies literate and preliterate. Class can be viewed in two quite different ways: First, it may be seen as a phenomenon that affects the magnitude and form of environmental heterogeneity and in particular tends to stratify it. This is of special importance because many of the environmental variables associated with class are likely to be important behavioral determinants. Class however, can also be considered as a component of population structure, superimposing vertical population units on geographical units since members of the same class typically tend to intermarry. There is of course considerable movement across the class boundaries through marriage and through social mobility and the nature and amount of this movement is of considerable significance to the behavior geneticist. Random flow between population units tends to be genetically homogenizing of the units. The flow that is selective, as vertical flow tends to be for at least some behavioral characteristics, produces stratification of any genes responsible for such behaviors. Typically selective flow will reinforce the effects of environmental variables so a so-called upper class will derive both the most desired genes and the most desired environments. One need say no more to demonstrate justification for knowledge of class structures in any genetic investigation or behavioral investigation at the population level.

Some of these general types of research questions can be well exemplified by studies my colleagues and I have recently been concerned with in Oxfordshire populations. Most of our attention has been directed to examining aspects of the population structure of a small group of rural villages but we have also paid some attention to the situation in Oxford city. (Boyce, Harrison, and Küchemann 1971; Boyce, Küchemann, and Harrison 1967, 1968; Harrison et al. 1974b; Harrison, Hiorns, and Küchemann 1970; Hiorns et al. 1969; Küchemann, Boyce, and Harrison 1967, 1971; Küchemann et al. 1974). The villages are mainly of Anglo-Saxon foundation and from the late sixteenth century have increasingly complete Parish records of baptism, marriage, and burial. Data are also available from the beginning of the nineteenth century from the national census returns. Linkage of the Parish records permits individual life histories to be established and families reconstituted, so it has been possible to follow changing patterns of fertility and mortality for about four centuries and to relate these to prevailing economic and social conditions.

Of special interest to my present concern are data from marriage registers that allow one to estimate the amount of genetic exchange which has occurred over time between the villages and between each of them and the outside world. It would appear from comparison of the marriage records with the census data that until quite recently most of the movement into and out of the villages has been associated with marriage. The amount of village exogamy has tended to increase linearly with time, but marital distances remained fairly constant until the mid-nineteenth century at around six to eight miles. However, with the arrival of the railway in the area these distances increased quite dramatically. On the basis of the observed marital exchanges it is possible, using a migration matrix approach, to predict the amount of genetic heterogeneity one would expect to find between the present-day populations of the different villages. As it happens, marital exchange has been sufficiently high to have brought the villages from states of initial heterogeneity to essential homogeneity within the period encompassed by the records, if movement were the only factor operating. But this homogeneity is mainly achieved through exchanges with the outside world and one can conceptualize the genetic history of villages as being one of being flushed through with genes from outside the area. In any event the prediction is one of genetic uniformity among the villages.

Marriage records from 1837 onwards contain information on the occupation of the groom, the groom's father and the bride's father. These occupations can be translated into social class categories using the Registrar General's (British office of vital statistics) key. Such categories then allow not only descriptions of the social class distributions in a population and the relationship between class and marital distance, but also afford measures of the amount of vertical movement through marriage between the classes and the social mobility of sons with respect to their fathers. Even between the villages, none of which have ever contained more than a thousand inhabitants, there is considerable variation in the frequency of the different classes, with the smaller villages tending to have a different pattern from the larger ones throughout the historical period for which there is information. What is more, even members of the same class tend to behave differently in terms of the amount of village exogamy and their preferences for mates of the same or a different class according to the village's size. The classes are themselves differentiated according to their preferences for intraclass marriages. For example, while persons in Class II and IV tend to marry within their own class, those in the intermediate Class III marry essentially at random with respect to all three classes.

There are, as one would expect, marked differences in the social class distributions between the villages and neighbouring Oxford City, and, again not surprisingly, different parts of the city are differentiated from one another. In the city there is and has always been since 1837 a striking relationship between class and marital distance. While this is not unexpected, the linear form of the relationship is remarkable with mean marital distance rising steadily from Class V to Class I. It is also noteworthy that the class of a bride appears to be as influential in determining this distance as the class of a groom. Clearly the ways genetic variety is dispersed by marital movement is a function of the class structure of the population.

The patterns of movement between the classes and thus the patterns by which relatedness between them can be expected to develop varies according to historical period and whether the population is an urban or rural one. Different urban groups are also distinguishable in these patterns. In general, marriage tends to be relatively more homogenizing in the village populations than social mobility, whereas the converse is the case in Oxford city. However, the class boundaries for both types of movement are less discrete in the urban group and there is more exchange across them in the twentieth century as compared with the nineteenth century. In both areas and both periods the vertical flow can be judged to be greater than the spatial flow, so even though the different classes are dispersing their genes over different distances this should not of itself lead to geographical variation in class differences. Indeed, the general prediction from the analysis of the movement between the social classes, both in Oxford city and the rural villages, is that there should be no genetic heterogeneity between the classes for those genes that do not affect the probability of movement. However, as has already been intimated this may not be true for any genetic basis to behavioral variation, since at least some of this behavior does influence choice of marriage partners and social mobility, in which case the movement will be stratifying rather than homogenizing in effect.

We have recently undertaken a human biological investigation of the rural village populations. This has involved a house to house survey in which people between the ages of eighteen and seventy were asked to complete a questionnaire concerning their educational, occupational, and movement experiences, and preferred handedness, to undertake parts of the Wechsler Adult IQ test, and the Eysenck Personality Inventory, to donate a blood sample for genetic and biochemical analysis, and they were measured for a series of anthropometric characteristics. The results are currently being analyzed, but it is now possible to make some broad descriptive statements about the populations. On the whole the genetic predictions as they concern blood groups and similar polymorphic markers are fulfilled and there is little if any genetic variation in these systems among the different villages and the different social classes. The personality characteristics, although admittedly very grossly surveyed, show also no geographical heterogeneity, but there are social class differences in measured levels of introversion. And when one comes to the IQ variation there is both marked social class and geographical variation.

Unfortunately, the composition of the populations has changed quite dramatically in the postwar years with considerable in-migration of urban commuters and out-migration of local people. Only in one village do there remain a substantitive number of locally born adults. Much of the pattern of IQ variation between the villages, both in verbal and performance ability, arise from varying proportions of immigrants of the different social classes. As a whole the immigrants have markedly higher IQ's. This, however, is no doubt largely due to selective migration for class, but even among peoples of the same class there is evidence that immigrants score higher than locally born people.

The main problem in interpreting the above results rigorously is that no information is available on locally born villagers who have left the area. They must obviously be likewise affected by the occupational opportunities provided in the essentially rural economy. As yet we have not examined the situation within families, but it is apparent that an attempt will have to be made to locate and test sibs who have left the area if even the family data are going to be meaningful. Just how fine some of the differentiations can be is indicated by a preliminary analysis of the patterns of assortative marriage for IQ. When this occurs it is almost entirely in the verbal component of IQ with little correlation between spouses for performance. However, even for the verbal element evidence for assortative marriage only exists when both spouses are locally born or nonlocally born. In the case of marriages involving an insider and an outsider there is no correlation irrespective of whether it is the male or female who is the outsider. We hope to be able to offer some explanation for why this should be the case, since we have information, as yet unanalyzed, on how the marriages came to be contracted, but this initial analysis indicates how dangerous it can be to generalize over even small populations.

Since, although for different purposes, information has been collected on both serological genetic markers and IQ in the same individuals, the data are being analyzed for possible associations between these characters. The rationale for such analysis has been discussed by J.M. Thoday (1967). Evidence has already been presented (Gibson et al. 1973) that in this Oxfordshire population an association exists between both the verbal and performance components of IQ and ABO blood group status, and indications of some other associations are being followed up. In the case of the ABO system, individuals of Group O achieved higher IQ scores than those of Group A, but the situation is complicated by the fact that the phenomenon essentially appears only in the locally born males and nonlocally born females. Selective migration could be responsible for this sex difference, which does not invalidate a conclusion that ABO status and IQ are associated. Such associations can arise in principle from pleiotropy or linkage disequilibrium and in the present case one would suspect that the latter explanation is the more likely. The association in the Oxford villages may therefore have no predictive value for other population, but if it is not artifactual, it does help in the more precise location of a genetic determinant for IQ.

Such findings are relevant to the general question of analyzing behavioral variety of the IQ kind between human populations. It has been argued most cogently that the usual methods of biometric analysis for the relative contribution of genetic and environmental variance are inappropriate for examining behavioral differences between groups that are physically identifiable. To this view I fully suscribe, if only because of the fundamental requirement of any scientific analysis that there is a need to randomize the effects of other determinants of the character examined against the determinant singled out for study. This, of course, is the essence of a scientific control and it is evident that in the case of so-called "racial" differences it is quite impossible to randomize cultural determinants of the behavior against possible genetic ones.

There thus seems little value in applying traditional within-population approaches to between-population differences and it has been said that an examination of the latter should be postponed until more is known of biochemical differences affecting behavior patterns. Such differences are more likely to be free of environmental interaction and could well be amenable to Mendelian analysis. While there is much to be said on scientific grounds alone for such postponement, interim results of value could be obtained by searching for associations between the polymorphic markers and gene determining quantitative behavioral variation in hybrid populations.

Where populations differ strikingly in the frequency of markers, and especially where genes present in one parental group are absent or rare in another an approach by establishing within-family associations is powerful in establishing linkage relationships. For this purpose, however, the hybrid groups need to be of recent but not too recent origin—ones in which it is possible to identify individuals and families of being of specific hybrid composition, that is, F_1, F_2, and first and second generation backcrosses, and in which linkage groups have not been totally broken up by crossing over. There appear, unfortunately, to be very few such groups. One was in the process of forming in various parts of Britain between West Africans and English a few years ago but changes in immigration pattern with the entry of many West Indians, whose hybrid ancestry is relatively long extant and not precisely knowable at the individual level, interrupted the process by providing the Afro/English hybrids with a much wider array of mates. Elsewhere, where primary miscegenation is occurring there also tend to be already in existence longer extant hybrids. Currently Indian/English hybrids are forming in Britain, but the process has not continued sufficiently long to provide the requisite number of generations and the same must be true elsewhere where hybrids are forming for the first time. However, for those interested in analyzing between population quantitative variety in behavior and who cannot await discoveries in neural-biochemistry it would seem more fruitful to attempt to locate the appropriate hybrid groups for association analysis than enter the polemic that inevitably arises in attempts to partition environmental and genetic components as they affect population differences and that is scientifically unresolvable.

References

Boyce, A.J.; Harrison, G.A.; and Küchemann, C.F. 1971. Population structure and movement patterns. In *Biological aspects of demography*, ed. W. Brass. London: Taylor and Francis.

Boyce, A.J.; Küchemann, C.F.; and Harrison, G.A. 1967. Neighborhood knowledge and the distribution of marriage distance. *Annuals of Human Genetics* 30: 335-38.

──────. 1968. The reconstruction of historical movement patterns. In *Record linkage in Medicine*, ed. E.D. Acheson. Livingstone: Edinburgh.

Gibson, J.B.; Harrison, G.A.; Clarke, V.A.; and Hiorns, R.W. 1973. IQ and ABO blood groups. *Nature* 246: 498-500.

Harrison, G.A. 1966. Human adaptability with reference to the IBP proposals for high-altitude research. In *The biology of human adaptability*, eds. P.T. Baker and J.S. Weiner. Oxford: Claredon Press.

Harrison, G.A.; Gibson, J.B.; Hiorns, R.W.; Wigley, J.M.; Hancock, C.; Freeman, C.A.; Küchemann, C.F.; Macbeth, H.M.; Saatcioglu, A.; and Carrivick, P.J. 1974b. Psychometric, personality and anthropometric variation in a group of Oxfordshire villages. *Annals of Human Biology*. In press.

Harrison, G.A.; Hiorns, R.W.; and Kücheman, C.F. 1970. Social class relatedness in some Oxfordshire parishes. *Journal of Biosocial Science* 2: 71-80.

──────. 1971. Social class and marriage patterns in some Oxfordshire populations. *Journal of Biosocial Science* 3: 1-12.

Harrison, G.A.; Küchemann, C.F.; Hiorns, R.W.; and Carrivick, P.J. 1974a. Social mobility, assortative marriage and their interrelationships with marital distance and age in Oxford City. *Annals of Human Biology* 2: 211-23.

Hiorns, R.W.; Harrison, G.A.; Boyce, A.J.; and Küchemann, C.F. 1969. A mathematical analysis of the effects of movement on the relatedness between populations. *Annals of Human Genetics* 32: 237-50.

Küchemann, C.F.; Boyce, A.J.; and Harrison, G.A. 1967. A demographic and genetic study of a group of Oxfordshire villages. *Human Biology* 39: 251-76.

Küchemann, C.F.; Harrison, G.A. Hiorns, R.W.; and Carrivick, P.J. 1974. Social Class and Marital Distance in Oxford City. *Annals of Human Biology* 1: 13-27.

Thoday, J.M. 1967. New insights into continous variation. In *Proceedings of the Third International Congress of Human Genetics,* eds. J.F. Crow and J.V. Neel. Baltimore: Johns Hopkins Press.

Commentary I

Sandra Scarr-Salapatek
University of Minnesota

Professor Harrison has offered several examples in this chapter of how population genetic ideas can be tested with anthropological methods. There is no reason, I agree, that cultural and social anthropologists should not be convinced to collect information in ways that test evolutionary and genetic hypotheses; family units are the minimum requirement; ethologically meaningful behaviors would be an ideal to which we should aspire.

Developmental psychologists should also be proselytized to design their studies to include genetically related and unrelated people of known environmental relatedness. The family parameter would add clarity to the results of all developmental studies in which individual variation is noticeable—in other words, in all developmental research. While general laws of behavioral development are useful abstractions, we tend to forget what Dobzhansky (1962) taught us long ago: that individuals are the real data to be accounted for. There is no necessary conflict between the description of average trends and the study of variations on that theme. There are different answers to the questions "How?" and "Why?"

Evolutionary theory offers an embarrassing wealth of ideas for the study of man's development, a wealth because the ideas are so many and varied, embarrassing because we have pursued so few. Mendelian genetic research is well advanced in the use of pedigrees and biochemical techniques to investigate aberrant development, but there has been little research on normal development. Ethologists have barely begun to apply their careful observational techniques and functional analyses to human behavior. Population genetics offers models of adaptation and variation for the developmental patterns of diverse peoples, but there are few data to subsume. Biometrical models, borrowed from animal breeding, have been applied (without sufficient modification)[a] to human populations for whom the necessary family data are scant and inadequate. Evolutionary biology,

---

[a]Since the major purpose of this brief discussion is to point out what can be done with methods and models we already have, I will not dwell on the shortcomings of biometrical models. There are serious problems in defining within- and between-family environmental terms for pairs of people of different genetic relatedness. For example, the within-family environments of twins are likely to be more similar than those of sibs by virtue of birth order, family composition, attitudes and expectations of others. The sources of variance in the environment need to be as carefully specified as the genetic terms in the same equations. Co-variances and interactions of genotypes with environments are also inadequately treated. The direct transfer of models from animal breeding (where environmental sources of variance are better controlled) to the human case is not acceptable.

particularly the theoretical efforts of Waddington (1968), E. Mayr (1970), and Dobzhansky (1970), should bring to developmental psychology a host of helpful ideas about the nature of man's development. Piaget (1970) and Werner (1948) recognized their influences, but little research has followed these leads. Finally, behavioral genetics, as it has evolved in the last fifteen years, offers to developmental research some ideas about the nature of behavioral plasticity, the choice of phenotypes for study, and the role of experience in phenotypic development.

Developmental studies can draw on the ideas and methods from several branches of evolutionary theory. We need not be embarrassed about eclecticism: The common theoretical thread is evolutionary and the common subject matter is development. The many questions about developmental phenomena require diverse concepts and methods.

There are elegant models in search of data. One primary task of the next decade should be to apply what methods we already have to test these models. Few populations in the world, as Professor Harrison notes, have been studied from an evolutionary perspective. Even fewer studies of any population have been framed with evolution and development simultaneously in the investigator's mind.

Studies of Populations

The study of breeding populations can refine genetic models and illuminate the effects of genetic and environmental differences. The parameters of population structure, such as geographic and social mobility, assortative mating, rates of reproduction, and the differential fertility of subgroups, are important elements in genetic models of breeding groups. How they actually combine to affect the distribution of behavioral characteristics in a population is something of a mystery. We have virtually no data on important groups such as United States blacks. Is geographic mobility related to any behavioral characteristics; is social mobility related to intellectual achievement, as it is in the white population; who mates with whom, with how many offspring? I submit that we know virtually nothing about any of this.

Only recently (Haseman and Elston 1972) have genetic models of linkage been tested on a few populations for a few behavioral traits. Further applications of the same techniques to better family data on more populations would answer questions about whether the linkage already suggested is general or whether it is only a local evolutionary adaptation, or an accidental association through mutation, or a pleiotropic effect of a single locus. New linkages may be discovered.

Developmental similarities and differences among genetically diverse and environmentally different peoples is of particular interest. There seem to be some robust developmental characteristics, such as acquiring speech, sensorimotor skills, right-handedness, and attraction to the opposite sex, that have particularly uniform patterns among populations, inspite of enormous apparent genetic and environmental diversity. These behaviors contrast with others like reading skills, formal operational thought, running speed, and sexual practices that can be quite different among groups. Might there not be some usefulness in applying evolutionary principles to the analysis of these uniformities and differences among populations? First, one would want to predict the minimal pattern of development for species membership. Second, of course, one would have to describe the developmental patterns for these characteristics in diverse populations. Then, one might be able to fit the model to the data. Darwinian methods for the study of subspeciation may well have some application to man.

Cross-cultural research of the usual type confounds culture with gene pool since cultural boundaries often define breeding populations. The simple description of existing behavioral differences between populations does not inform us of how and why these differences arose. To attribute them to conveniently describable environmental or genetic differences only begs the issue. There is no simple way to disentangle the origins of behavioral differences in the usual cross-cultural study (Pick in press). A theoretical approach to population studies that predicted behavioral similarities and differences would shed quite a new light on cross-cultural research.

Variations within Populations

Family studies are the method of choice for the investigation of developmental variations within a population. There are two seldom-used family constellations that have particular promise: the offspring of monozygotic (MZ) twins and adoptive families. All family studies are logistically difficult and expensive, but there is little alternative if one wishes to answer questions about true environmental effects, the effects of genotypes with environments, and developmental plasticity.

A valuable and economical constellation of family members is provided by the families of adult MZ twins. In the two nuclear families of the adult twins one finds the following pairs of related people: (1) monozygotic twins reared together (MZT), (2) parents and offspring reared together (POT), (3) parents and offspring reared apart (POA), (4) siblings reared together (ST) and half-siblings reared apart (HSA). Figure 4-1 gives the family constellation.

By studying a minimum of eight people per MZ twin family constella-

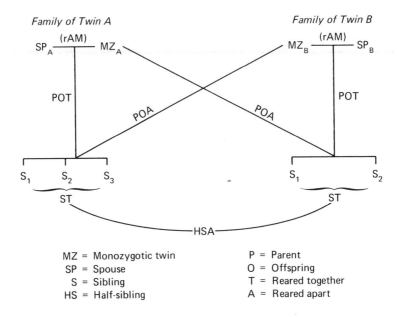

Figure 4-1. The Genetic Relatedness of the Families of Monozygotic Twins

tion (four parents and at least four children) the investigator obtains a data set that would satisfy the most demanding biometrician (Jinks and Fulker 1970). The total sample of families, however, might need to be large—in the hundreds—to satisfy statistical requirements (Eaves 1972). Thus, a collaborative effort among investigators in various areas of the country is probably required. Dr. Karen Fischer has already begun a small-scale study to test the feasibility of the approach.

From these same data one can calculate an assortative mating coefficient (rAM), regressions of midparent on midchild values, and maternal versus paternal effects on development. Assortative mating for any measured characteristic is the correlation between spouses. In the case of MZ twins, one might also be interested in the correlation between their respective spouses as an estimate of the degree of environmental similarity chosen by the same genotype on two occasions.

The adoption family design offers still other advantages, particularly for the study of environmental effects. The behavioral stratification of populations by socioeconomic characteristics presents difficult problems of interpretation when children are reared by their own parents. Not only do they share a family environment, but they are genetically related, and the genes and environments are correlated. In adoptive families shared environments are distinct from shared genes. To the extent that biological stratification exists, adopted children's behavior should correlate with their biological parents. To the extent that social environments rank order children's behaviors, adopted children's scores should correlate with their adoptive parents. The adoptive family study is the best method for studying true environmental effects.

Developmental plasticity can also be evaluated from adoptive family data. Predictions can be made about children's scores from their biological parent's values. The level of the children's scores can be compared with predictions from biological parents and with adoptive parents' scores. There will doubtless be much debate over interpretations of such data based on possible underestimates of the biological parents' "genetic values" and selective adoption procedures that eliminate abnormal children from the pool of adoptees. Both underestimating the biological parents' "genotypes" (because their phenotypes developed in poor environments) and selecting children only in the normal range will tend to exaggerate the difference between predicted levels and actual levels of performance for adopted children.

Some would argue that intervention studies provide a better estimate of developmental plasticity. Indeed, they are more straightforward than adoptive families in their attempts to increase performance levels. On the other hand, educational programs seldom continue across a child's development from the first year to maturity, and they seldom offer as intensive or extensive an intervention as a family environment. Only studies of kibbuzim and adoptive families seem to offer these advantages. Finally, educational manipulations seem to lack an intuitive kind of ecological validity; they are contrived, limited experiments. I think we will learn more about the extent of developmental plasticity for many characteristics from adoptive families than from educational experiments (which have other values, however).

An additional use of adoptive family data is to study genotype × environment interactions (Vge). In the strict sense of Vge, only monozygotic twins reared apart offer the opportunity to study the *same* human genotype reared in two environments. To the extent that the environments of co-twins are uncorrelated, one could estimate the differential effects of environments on genotypes. But monozygotic twins reared apart are rare creatures, not likely to be available in large enough numbers for meaningful studies of Vge. Besides, who knows how selected their genotypes are; who gives up twins for adoption in separate families?

A more economical design involves average predicted values for groups of adopted children, based on their biological parents' scores.[b] If the

---

[b]If biological parents' socioeconomic characteristics correlate with those of the adoptive parents, selective placement will have to be partialled out of the correlations between adoptive parents and their children. Unlike natural families, however, the degree of genetic correlation in adoptive families is likely to be small, and it can be calculated.

children are nearly randomly assorted into adoptive families of varying characteristics, one could examine the effects of similar adoptive family values on children with different average predicted outcomes and the effects of different adoptive family values on children with similar predicted outcomes. The analogy is to different genotypes in the same environment and to the same genotypes in different environments. If the children of bright natural parents, for example, score equally well regardless of adoptive family characteristics but duller children score well only in families with bright adoptive parents, this would constitute one form of genotype by environment interaction.

Developmental Behavior Genetics

Development studies require more than behavior genetic research usually offers. The developmental questions refer to ontogeny, to changes over time in the organization of behavior. The perspective is the life span. Behavior genetic research can illuminate the origins of developmental shifts. A developmental perspective can uncover genetic timing mechanisms that might not be noticed in cross-sectional research.

There are interesting genetic questions to be asked about the invariant sequences in language acquisition and cognitive development, to mention only two. Individual variation in patterns of cognitive development have received some attention (see Scarr-Salapatek, in press, for a review). Studies of variation in the timing and pattern of language acquisition have hardly begun. The developmental patterns themselves suggest maturational timing mechanisms that are species specific. This is not a new idea (Piaget 1970), but we do not know what the mechanisms are.

The Schaie longitudinal-cross-sectional design is particularly appropriate for developmental behavior genetic studies (see Schaie 1965 and Chapter 9, this book). Cohorts of families, instead of individuals, could be chosen and followed. Studies at behavioral and biochemical levels should be coordinated to look for genetic timing mechanisms and sources of individual variation in the same subjects.

Behavior genetics is already an interdisciplinary field. Collaborations among geneticists, biochemists, and psychologists are frequent and well established. To add a more developmental flavor to the field would seem to be a relatively simple matter: Developmental psychologists and behavioral geneticists need only to be introduced to each other's questions and techniques. Then we shall see more investigators with development and evolution simultaneously in mind.

References

Dobzhansky, T. 1962. *Mankind evolving*. New Haven: Yale.

―――. 1970. *Genetics of the evolutionary process*. New York: Columbia.

Eaves, L.J. 1972. Computer simulation of sample size and experimental design in human psychogenetics. *Psychological Bulletin* 77: 144-52.

Haseman, J.K., and Elston, R.C. 1972. The investigation of linkage between a quantitative trait and a marker locus. *Behavior Genetics* 2: 3-19.

Jinks, J.L., and Fulker, D.W. 1970. Comparison of the biometrical, genetical, MAVA, and classical approaches to the analysis of human behavior. *Psychological Bulletin* 73: 311-49.

Mayr, E. 1970. *Populations, species and evolution*. Cambridge: Harvard.

Piaget, J. 1970. Piaget's theory. In *Carmichael's manual of child psychology*, ed. P. Mussen, vol. 1, pp. 703-32. New York: Wiley.

Pick, A. In press. Culture and perception. In *The handbook of perception*, eds. E.C. Carterette and M.P. Friedman. New York: Academic Press.

Scarr-Salapatek, S. In press. An evolutionary perspective on infant intelligence: Species patterns and individual variation. In *Infant intelligence*, ed. M. Lewis. New York: Plenum.

Schaie, K.W. 1965. A general model for the study of developmental problems. *Psychological Bulletin* 64: 92-107.

Waddington, C.H., ed. 1968-72. *Towards a theoretical biology*, vols. 1-4. Chicago: Aldine.

Werner, H. 1948. *The comparative psychology of mental development*. New York: International Universities Press.

Commentary II

Lissy F. Jarvik
University of California at Los
Angeles and Brentwood V.A.
Hospital

To someone who has spent many years in the quest for knowledge of genetic influences upon human behavior, Dr. Harrison's opening statement in this chapter is like the proverbial gauntlet thrown to the medieval knight. Worse yet, there is no defense; Dr. Harrison is quite correct; with respect to behavioral variety, there *is* "very little information on most of the world's populations that is likely to be of value to geneticists." Worst of all, Dr. Harrison's cautions regarding research design and population selection, cautions well taken and often disregarded in past studies, generate general pessimism concerning the feasibility of valid research in the area of human behavior genetics. Dr. Harrison leaves us but two options: (1) To wait "until more is known of biochemical differences affecting behavior patterns," or (2) "for those interested in analyzing between population quantitative variety in behavior and who cannot await discoveries in neural-biochemistry it would seem more fruitful to attempt to locate the appropriate hybrid groups for association analysis than enter the polemic which inevitably arises in attempts to partition environmental and genetic components as they affect population differences and that is scientifically unresolvable."

Fortunately, Dr. Harrison has not heeded that advice or we would have been deprived of the fascinating information in this chapter as well as in his previous papers. Rather than discuss the substantive issues raised by his research findings, I shall address my comments to certain more peripheral but important aspects of his discussion, such as his plea for multigenerational family studies, a plea that I should like to underline and that should be heeded not only by anthropologists, but by other behavioral scientists as well, and not by behavioral scientists alone, but by biochemists and clinicians too.

Dr. Harrison draws our attention to the fact that such family studies "demand not only a much greater involvement and commitment on the part of investigators, but also a high level of cooperativeness from potential subjects." Having been initiated into family studies more than twenty-five years ago by the late Franz Kallmann and the late Gerhard Sander, I can attest to the accuracy of Dr. Harrison's statement. There is another factor to consider in longitudinal family studies and that is the finite life span of the investigators. In our own study of aging, many of the subjects have outlived the first generation of investigators, and these subjects were already in the seventh decade of their life when they entered the study (Blum, Jarvik, and Clark 1970; Jarvik et al. 1962; Kallmann and Sander 1948). Moreover, this was essentially a single-generation study! Nonetheless, if such studies are properly planned and adequately financed with long-term support guaranteed, they are feasible.

Dr. Harrison further remarks that cooperativeness is generally higher among rural than among urban groups and, of course, mobility is generally greater among the latter, thus complicating research and raising the question of general representativeness of population groups consisting of rural residents exclusively. True, they are not representative of the population-at-large. However, they may well be suitable for preliminary investigations. We must ask ourselves whether, at this stage of our knowledge, we need all of the sophistication in research design and population sampling available to us. Since we know so little, most of our research in behavior genetics has the characteristics of "fishing expeditions," in the sense that we are trying to detect relationships rather than explore the characteristics and general applicability of relationships already discovered. Even if we work with a group that is not generally representative, the detected relationship will be true for that particular group, and we can proceed thereafter to test it on other populations. The situation is reminiscent in part of that in drug studies where it is really unnecessary to use double-blind, well-controlled procedures until some evidence has been gathered in open studies for the effectiveness of a given drug. If a drug is found to be ineffective in open trial, then the need for more sophisticated research techniques no longer exists. In behavior genetics, too, we might first try to describe relationships in limited populations, knowing full well that we may discover relationships that are generally not important or, in turn, that we may miss relationships that are important. Nonetheless, it would seem to be an economically sound starting point.

To return to Dr. Harrison's comments, I should like to take up his suggestion that we either wait for discoveries in neurobiochemistry or rely on association analysis in appropriate hybrid groups. I do not think that we have to restrict ourselves to these alternatives. We could use genetic information to change directions entirely. For example, we need not limit ourselves to the use of twins for the partitioning of genetic and environmental variances. We could, instead, design experiments to take advantage of the fact that in monozygotic twins genotype is essentially constant rather than variable. In order to do so we need to define our goals more carefully than we have tended to do so far. For example, most of the research on genetic aspects of intellectual functioning has been concerned with IQ. Dr. William Meredith (during the discussion of Dr. Harrison's paper at the workshop which led to this book) went so far as to question the importance of both general intelligence (IQ) and height. Having thought about this statement ever since he made it, I have been unable to arrive at any attribute that would approximate in importance either general intelligence or size, whether in terms of value for survival or for day-to-day functioning. There are few, I believe, who would disagree that among the features distinguishing homo sapiens from lower species, intelligence ranks high—we need merely translate the scientific designation of human beings. When it comes to size, however, we generally regard ourselves as puny; and so we are—compared with the elephant, the tiger, the lion, and the giraffe. We rarely consider that "99% of animal species are smaller than we are. Within our own order of approximately 190 species, only the gorilla regularly exceeds us in size" (Gould 1974).

Regardless of the importance that we may, or may not, assign to size, the major value of the IQ and, in my opinion, its only proven value at this time lies in its ability to predict scholastic performance. When we attempt to change IQ, we are really interested in changing not the IQ itself but rather the scholastic performance. As mentioned earlier, Dr. Meredith suggested that we abandon IQ and pay attention instead to tests of spatial abilities. Indeed, the latter have been generally neglected. However, I would plead for addition, not substitution. Dr. Henderson in his able review summarized the case for substituting specific factors for IQ scores and/or contrasting fluid with crystallized intelligence. However, until we can specify the meaning of the various factor scores, or of fluid and crystallized intelligence, in terms of the specific achievements that we consider important, changing from IQ to other measures of intellectual functioning may not be as profitable as the advocates hope at present. I have no objection to pursuing these lines of investigation, but I fear that we may create the same dilemma of attempting to partition inextricably interwoven interactions between heredity and environment. We hope that more powerful techniques will be devised that will enable us to assess such interactions.

Dr. Meredith's comments regarding the omission of spatial abilities from tests of general intelligence may be misunderstood in the sense that such omissions were due to oversight. They were not. On the contrary, the omission has been deliberate, and everyone who has attempted to construct an intelligence test in such a manner that the scores of boys and girls or men and women would equal each other has found it impossible to include measures of spatial abilities since boys and men almost invariably outscore girls and women on such tasks. There is suggestive evidence for a sex-linked recessive mode of inheritance for some factor or factors concerned with spatial ability or its suppression (Bock and Kolakowski 1973; Corah 1965; Guttman 1974; Hartlage 1970; Stafford 1961). In addition, there appear to be potent hormonal influences required early in life, probably in the fetal stage, for the development of good spatial ability (Money

1971). Considerable further research is needed in this area, but in the meantime we must be careful not to rush in and substitute spatial ability for IQ or we may find that this substitute, like most substitutes, is far less satisfactory than the original. Particularly, the implication that spatial abilities are as good or even better a measure of a human being's creativity, genius, and contribution to society must be examined with care and substantiated by a solid body of acceptable evidence. Otherwise, the inevitable classification of the second sex as intellectually inferior will once again gain acceptance based on personal credo and the authoritative stature of its proponents rather than upon any scientific grounds.

Instead of merely substituting one set of measures for another, a more productive approach might be to set definitive goals, for example, to teach children of a given age to read with a predetermined level of proficiency after a predetermined period of instruction. There should be no difficulty in accurately, reliably, and validly measuring the preinstructional and postinstructional reading levels and in equating reasonably well the type of instruction administered. It should be possible to get several groups composed of monozygotic twin pairs, each group being homogeneous for the level and patterns of intellectual functioning, as measured by factor scores or otherwise. Different teaching methods could then be applied to the two halves of each group, each half being composed of one of the two co-twins. It should not take too long to determine in that way which method of instruction appears to be best suited for which level and pattern of intellectual functioning.

Such an experimental design has three advantages: (1) Twins are readily available, and (2) the studies could, therefore, be carried out at minimal expense, and (3) such studies would be entirely ethical since there is no clear-cut evidence at this time in favor of one teaching method over any other. And, yet, that approach has generally not been used. I cannot pretend to be original in proposing the design since it was Gesell who did so some thirty-three years ago. Nor is it unknown in the psychological literature, Gesell and his students having published extensively on the co-twin control method (Gesell and Thompson 1941). Yet, if the method has so many advantages and no apparent disadvantages, why then is it not used? There is no ready explanation except, perhaps, when it comes to *human* behavior, *anything* to do with genetics is taboo; even if genetic knowledge is used merely to reduce the number of variables.

Although the most desirable use of the co-twin control method is in setting up prospective studies, a variant of it can be used retrospectively as well. Thus, in our own study of intellectual changes with advancing chronological age, we discovered a number of monozygotic pairs discordant for a specified degree and rate of cognitive decline, termed "critical loss" (Jarvik and Blum 1971). Since critical loss has been demonstrated as a powerful predictor of individual morbidity and mortality (Blum, Clark, and

Jarvik 1973) we know that the critical loss as well as the highly correlated morbidity and mortality are significantly influenced by environmental factors. If we hypothesize, as we have done earlier, that critical loss is a reflection of subclinical atherosclerotic changes, then the discordance of monozygotic pairs should serve as a potent stimulus for further search for critical environmental factors responsible for their discordance (Jarvik, Blum, and Varma 1972). So far, unfortunately, we do not have the answer from our own material and, as far as I know, no one else has the answer either. It should be feasible to collect a series of such discordant pairs with identical genotypes and they would constitute an ideal group for control trials of dietary and other therapeutic regimens. After all, the discordant twin without critical loss would provide an estimate of the lower limit of potential survival within that particular genotype. Effectiveness of a variety of intervention programs could be judged by how closely the life span and health of the twin with the critical loss could be brought to approximate that of the discordant co-twin. These are but isolated examples of how twin studies could be utilized in novel ways and our imaginations alone set limits to their usefulness.

An entirely different approach has come to mind as a result of the paper recently published by DeFries et al. (1974). A battery of fifteen cognitive tests was administered in Hawaii to residents of either Japanese or European ancestry, and the result "yielded the same four major cognitive factors for each of the two ethnic groups, and these factors are defined by strikingly similar factor loadings." Identifying certain patterns of cognitive functioning may be a more profitable way of comparing various ethnic groups than the usual resort to mean scores and variances. Indeed, I could easily conceive of extending such studies beyond homo sapiens and, utilizing the appropriate tests, attempt to identify as a start, similarities in cognitive structures between primates and humans and then ranging even beyond that. After all, it should not be surprising to find similarities in cognitive structures where we find similarities in physiology, pharmacology, and immunology, to name but a few. In chromosomal structure the resemblances between human and primate, and even nonprimate monkeys, easily outweigh the differences and the same may well be true of DNA content. We share many basic behaviors with other species and just as today we would not think of trying out a drug on a human being lest we had made extensive preliminary studies of its toxic as well as therapeutic effects in other organisms, so I could visualize the time at some future date when we would no more think of applying certain teaching methods to our own children lest we had tested their beneficial as well as noxious effects upon some other species. We tend to be very careful about our bodies and the way we manipulate them while we blithely disregard all cautions when it comes to our psyches.

I believe the time has come when, as scientists, it behooves us to shed our prejudices and preconceptions in order to deepen the understanding of our mental functioning, and the influences that determine it, to at least that extent which will parallel the understanding we have gained of our physiological functioning and the influences determining that. One way to start would be to use the tools already in our possession for the benefit of the next generation, however unpopular a course that may be today. After all, depending on the vantage point, a thousand years may seem as yesterday—when it is past (Psalm 90:4).

References

Blum, J.E.; Jarvik, L.F.; and Clark, E.T. 1970. Rate of change on selective tests of intelligence: A twenty-year longitudinal study of aging. *Journal of Gerontology* 25: 171-76.

Blum, J.E.; Clark, E.T.; and Jarvik, L.F. 1973. The New York State Psychiatric Institute study of aging twins. In *Intellectual functioning in adults*, eds. L.F. Jarvik, C. Eisdorfer, and J.E. Blum, pp. 13-19. New York: Springer.

Bock, R.D., and Kolakowski, D. 1973. Further evidence of sex-linked major-gene influence on human spatial visualizing ability. *American Journal of Human Genetics* 25: 1-14.

Corah, N.L. 1965. Differentiation in children and their parents. *Journal of Personality* 33: 300-308.

DeFries, J.C.; Vandenberg, S.G.; McClearn, G.E.; Kuse, A.R.; and Wilson, J.R. 1974. Near identity of cognitive structure in two ethnic groups. *Science* 183: 338-39.

Gesell, A., and Thompson, H. 1941. *Twins T and C from infancy to adolescence: A biogenetic study of individual differences by the method of co-twin control.* Provincetown: Journal Press.

Gould, S.J. 1974. This view of life: Sizing up human intelligence. *Natural History* 83: 10-14.

Guttman, R. 1974. Genetic analysis of analytic spatial ability—Raven's progressive Matrices. *Behavior Genetics*. In press.

Hartlage, L.C. 1970. Sex-linked inheritance of spatial ability. *Perceptual and Motor Skills* 13: 610.

Jarvik, L.F., and Blum, J.E. 1971. Cognitive declines as predictors of mortality in twin pairs: A twenty-year longitudinal study of aging. In *Prediction of life span*, eds. E. Palmore and F.C. Jeffers. New York: Heath.

Jarvik, L.F.; Blum, J.E.; and Varma, A.O. 1972. Genetic components and intellectual functioning during senescence: A 20-year study of aging twins. *Behavior Genetics* 2: 159-71.

Jarvik, L.F.; Kallmann, F.J.; Lorge, I.; and Falek, A. 1962. Longitudinal Study of Intellectual Changes in Senescent Twins. In *Social and psychological aspects of aging,* eds. C. Tibbits and W. Donahue. New York: Columbia University Press.

Kallmann, F.J., and Sander, G. 1948. Twin studies on aging and longevity. *Journal of Heredity* 39: 349-57.

Money, J. 1971. Prenatal hormones and intelligence: A possible relationship. *Impact of science on society* 21: 285-90.

Stafford, R.E. 1961. Sex differences in spatial visualization as evidence of sex-linked inheritance. *Perceptual and Motor Skills* 13: 428.

5

Genetic Mechanisms in Human Behavioral Development

Gilbert S. Omenn
University of Washington

Introduction

Poets share with geneticists a consuming interest in describing, if not explaining, the nature of human individuality. Many attributes serve to distinguish people in the eyes of other people; yet, it is clear that our behaviors are the basis of the key observations. At the same time, there is a compulsion to generalize—to describe what theologians call man's "likeness" to God; what poets call "the human experience," what scientists discern as "unity within diversity." For the purposes of organizing this chapter, let me cite four elements of the search for biochemical genetic mechanisms involved in the development of human behavior:

Gene Action in the Relevant Organs

Although the circulatory and respiratory systems and really the entire bodily functions are essential for support of the nervous system, we may restrict the discussion of organ systems primarily involved in behavior to the nervous system and the endocrine glands and their feedback loops. A tremendous body of evidence from many species indicates that there is a generality in the overall direction of gene action—that genetic information is coded in the DNA of cell nuclei (in RNA only in certain RNA viruses); that the genetic information flows from the DNA code via RNA messengers to protein products, with regulatory steps affecting the timing and magnitude of synthetic and degradative processes. All cells, including neurons, neuroglial cells, and hormone-producing cells, contain the same complement of DNA, but the regulatory processes of cellular and tissue differentiation lead to different patterns of gene activation in different tissues. Approaches to a description of these different patterns, including the timing and magnitude of transitions in the pattern of a particular developing tissue, are central to our discussion. The potential foci for study include

---

Preparation of this chapter was supported by Genetics Center Grant GM 15253 and by a Research Career Development Award from the United States Public Health Service.

DNA, RNA of various classes, proteins, complex proteins, lipids, membranes, as well as rates and interaction of metabolic pathways.

Molecular, Biochemical, and Metabolic Differences Among Normal Individuals

Examining the same kinds of biochemical parameters that may be investigated in studies of brain and endocrine tissue in general, these studies would focus on genetically determined variation in the properties of the RNA transcript of the DNA code, the protein products, including polymorphisms, and membrane receptor functions. The description of differences is only a beginning. Next, criteria must be applied to justify the conclusion that the differences are genetically determined. For example, a difference in quantitative activity of an enzyme could reflect many influences. If an electrophoretic difference is demonstrated for an enzyme prepared in identical fashion from two different individuals, it is likely that a structural difference (altered net charge due to amino acid substitution) is involved. To demonstrate the genetic basis, however, would require a family study. The subsequent step in this element of our discussion is still more difficult: correlation of identifiable biochemical differences with behavioral data. Associations of biochemical and behavioral data will be related more rationally if descriptions can be made at intermediate levels with physiological, electrophysiological, and psychological measures. It is obvious that biochemical studies on human brain are severely constrained by the lack of material, especially lack of tissues obtained from individuals on whom any meaningful behavioral data are available.

Differences among normal individuals may be revealed also under conditions of metabolic "stress." For example, the administration of pharmacologic agents with primary action on the nervous system may detect differing susceptibility of enzymes or membrane receptors to the agent and allow inferences about related differences in affinity for endogenous agents, such as neurotransmitters. Another kind of "probe" of the responsiveness of the nervous system is steroid hormones, particularly the sex steroids, and any other metabolic factors in the fetal and postnatal development of sex differences.

Mutant Analysis of Behavior

A growing array of enzyme deficiencies has been recognized as "experiments of nature" in man. Some of these are associated with mental retardation, some with other behavioral abnormalities; others affect only red blood cells or other tissues, and some seem to have no detectable deleterious effects. Most are inherited as autosomal recessive traits, though a few are X-linked recessive traits manifested in males in hemizygous form. Some of the disorders affecting behavior are due to toxic effects of metabolites accumulated as a result of metabolic defects in other tissues, while other disorders are intrinsic to the nervous system. Toxic mechanisms, as far as the brain is concerned, are quite analogous to environmental insults, including oxygen deprivation or lead intoxication. It will be important to analyze "normal variation" to seek explanations for the greater susceptibility of some individuals than others to the deleterious metabolic effects of such toxic agents. Another area of potential yield in this category is the large number of inherited disorders of the sense organs, particularly the visual and auditory apparatus. Relatively little attention has been paid to the effects on central nervous system functions of disordered sensory input or restricted effector capability in the muscular and autonomic systems.

Role of Inherited Biochemical Differences in the Cellular Organization of the Nervous System

Given the billions of cells in the nervous system and their myriad connections, their patterns of migration and selective cell death, it is fair to ask whether the specific cell-cell interactions of the nervous system represent the specific directives of a genetic programme or some more stochastic process that fits together the bits of genetically determined components. It is quite possible that we could not predict the cellular organization of the nervous system even with a complete description of the molecular and biochemical nature of its component enzymes, membranes, and cells. Many biochemical specificities may be induced intracellularly or at the cell surface only after certain interactions are initiated. This large question raises some analogies to the computer-simulated models of brain organization—relatively simple bits of information can be organized in increasingly complex patterns and similarly complex patterns may be generated by different specified or even random routes. I introduce this element to our discussion as a sobering note: Even if we were successful beyond our most optimistic expectations in describing gene action and genetically determined biochemical variation in the nervous system, we might be far from understanding how the system developed and functioned electrophysiologically and behaviorally. At such a point, however, genetic approaches would again prove most valuable, providing individuals or cultured organ systems of defined genetic relationship as a parameter with which to examine the predictability or unpredictability of nervous system organization.

Gene Action in the Nervous System

The "resting" state of neurons is characterized electrophysiologically by intense rhythmic and spontaneous activity, in striking contrast to the old view of a stable set of quiescent neurons that could be stimulated to action. The brain consumes up to 50 percent of the resting energy and oxygen supply of the body, yet constitutes only 2 to 3 percent of body weight. Even though cell division ceases early in life for neurons, protein biosynthesis and transport of proteins, structural components, and other molecules through the axon of neurons are continuous, highly active processes.

The importance of relating biochemical and metabolic findings to regions of the brain, to specific fiber connecting pathways, and to subcellular compartments cannot be overemphasized. Even more than other tissues, where the same considerations do apply, the brain has a complex organization that may mask significant local differences and may lead to discrepancies in the results of separate experiments or different laboratories. For example, the conditions for maximal protein synthesis may be coupled to specific neuronal functions (Shooter 1972). In large neurons the cytoplasmic ribosomes are concentrated near the endoplasmic reticulum of the Nissl substance in the perinuclear region, the initial segment, and the axon hillock (Sotelo and Palay 1968). A high proportion of these ribosomes are not attached to the membrane of the endoplasmic reticulum and may function directly in the synthesis of protein involved in axoplasmic transport to the nerve ending. Brain ribosomes require a high concentration of potassium ion (100 mM K^+), suggesting a link of protein biosynthesis with bioelectric phenomena and active transport of K^+. Mitochondrial protein synthesis is tightly linked to oxidative phosphorylation, increasing under conditions optimal for the latter and being inhibited when specific inhibitors of oxidative phosphorylation (rotenone or antimycin A) are present. Finally, nerve ending fractions called synaptosomes carry on protein synthesis. The ionic concentrations required for maximal incorporation of amino acids into protein also result in maximal sodium-potassium ATPase activity, potassium uptake, and oxygen uptake. Ouabain, which inhibits the Na-K ATPase activity, markedly inhibits synaptosomal protein synthesis, further suggesting a close coupling of the synthetic activity with the ionic flux and energy metabolism of the nerve ending.

One of the most dramatic findings of overall gene action in the nervous system has come from studies of *DNA-RNA hybridization*. Differentiation of tissues leads certain genes to be active in certain tissues, other genes to be active in other tissues, and some genes to be active in all tissues. DNA-DNA hybridization confirms that all normal cells of an organism contain the same DNA, derived from the DNA complement of the fertilized egg, while DNA-RNA hybridization confirms only that part of the genome is active in any tissue at any time (McCarthy and Hoyer 1964). Some genes are redundant, coding for large amounts of ribosomal RNA needed for protein synthesis. However, appropriate methods can determine the proportion of "unique sequence DNA" genes present in single copies and transcribed into RNA messengers that direct protein synthesis. In such tissues as liver, kidney, and spleen, only 3 to 6 percent of this DNA is transcribed into extractable RNA (Grouse, Chilton, and McCarthy 1972; Hahn and Laird 1971). In brain tissue a remarkably higher proportion is transcribed: 10 to 13 percent in the mouse brain and approximately 20 percent in human brain (Grouse, Omenn, and McCarthy 1973).

Our preliminary findings (Grouse, Omenn, and McCarthy 1973) suggest that the values for diversity of RNA messengers are substantially higher in cortex than in the brain stem. More tentative results indicate that dominant cortex may have higher values than nondominant and that normal appearing cortex has higher values than cortex that appears grossly atrophied. These findings are quite exciting, for a value of 20 percent hybridization means that 40 percent of the genome (transcribing from one strand *or* the other) is expressed. Since the amount of DNA per cell (about 6 picograms) is the same in the mouse and in man, these data strongly indicate that evolution has been associated with an increasing assignment of the genome to functions of the central nervous system, particularly the cortex. Such a biochemical finding is certainly consistent with anatomical and psychological characterizations of the development of higher cortical functions. Our limited data also suggest a much higher value for DNA-RNA hybridization in adults than in fetal brain. We have not ruled out the possibility that a detectable fraction of the genome is active in fetal brain and not active later. In fact, a great many experiments remain to be done in this area.

The experiments are extremely demanding technically. Highly purified DNA must be prepared and then fractionated to provide "unique sequences." The DNA must be radioactively labeled, either by incorporation of tritiated (^{3}H)-thymidine during DNA biosynthesis (in cultured cells) or by labeling purified preparations. We have been attempting to utilize DNA highly labeled with I^{125}. This approach has the advantages of allowing the preparative steps to proceed without introducing radioactivity and then yielding DNA with much higher specific activity. The I^{125}, of course, does decay with a half-life of sixty days. A greater variety of labeled DNA preparations should be possible with this method. The second component of this system is highly purified RNA extracted from the tissue under study. All DNA must be removed or degraded, since even a tiny fraction of DNA will greatly alter the hybridization results. If RNA is degraded, particularly if degradation occurs in vivo or postmortem, some effort is required to determine whether disproportional degradation over time of particular fractions of the RNA may have occurred. Such loss of RNA should diminish hybridization values, not increase them, so the general observation of high hybridization values in brain would hold. However, detailed comparisons from different specimens or different regions of brain might well be complicated by this factor. Once good preparations of DNA and of RNA are obtained, the hybridization reactions and the subsequent differentiation of single-stranded DNA from double-stranded DNA/RNA hybrids are highly sensitive to technical conditions.

Among the experiments on the drawing board are the following: (1) repetition of the reported studies, utilizing I^{125}-DNA in place of ^{3}H-DNA; (2) extension of comparisons of cortex with brain stem, dominant with nondominant cortex, adult with fetal brain; (3) comparison of hybridization values for various regions; (4) establishment of some estimates of the confidence limits for the extraction technique and for various regions of the brain; and (5) examination of the impact of certain pathological states. Parallel studies are being carried out with mouse and especially monkey brain. There is no cross hybridization between human and mouse DNA and RNA. However, cross hybridization can be obtained with human and macaque material; brain of chimpanzee will be of special interest. Also, it will be interesting to determine the general level of hybridization in the cortex of these nonhuman primates, with the expectation that the value will be somewhere between those obtained for the mouse and the human brain. Finally, one experiment of deliberate environmental modification can be performed: Dr. Norman Henderson has provided brains from two series of mice put through either a normal (relatively deprived) or a highly enriched cage environmental experience.

One other question already has been approached in our early studies. Since brain has highly differentiated cell types, it is conceivable that the very high hybridization values represent the additive effects of genomic expression in several different types of cells. We plan to set up addition and competition experiments eventually, but current conditions of hybridization require RNA concentrations already at the limits of solubility. Grouse, Omenn, and McCarthy (1973) did examine RNA extracts from two essentially clonal populations of nerve cells, the mouse neuroblastoma line cultured in vitro and human glial malignant tumors obtained at surgery. Mouse neuroblastoma RNA hybridized with 3.5 percent of unique-sequence mouse DNA, and RNA isolated from two human astrocytomas was complementary to about 9 percent of the human DNA: These values correspond to 30 to 40 percent of the values obtained with whole brain RNA extracts in the respective species. If the RNA populations extracted from tumor cells reflect transcription in that class of cell in the normal brain, then neuronal cells and glial cells appear to exhibit roughly the same extent of transcriptional diversity.

It is tempting, through probably premature, to speculate about what functions the greater diversity of RNA messengers may serve in brain cortex. Language and memory functions are particularly interesting possibilities, of course.

Thus far, we have concentrated this discussion upon the DNA to RNA limb of gene expression. Let us turn now to the protein products. Studies of the proteins of different tissues are of essentially two kinds. First, there are comparisons of the physical properties of proteins of unknown function, simply identified as separable "bands" on polyacrylamide electrophoretic gel systems, for example. Second, there are studies of particular enzymes or other proteins whose functions are known, preferably involving key metabolic steps.

Electrophoretic profiles of brain proteins appear significantly different from similar preparations of other tissues (Caplan, Cheung, and Omenn 1974). Two proteins highly specific for brain have been found: The S-100 protein (named for its solubility in 100 percent saturated ammonium sulfate) in glial cells (Moore, Perez, and Gehring 1968) and the 14-3-2 protein (named for its position on three successive chromatographic elutions) in neuronal cells (Cicero et al. 1970). The electrophoretic method has, indeed, been applied to seek developmental changes. Recent studies of the profiles of aqueous-soluble and insoluble proteins have demonstrated changes during development in whole brain extracts of the mouse (Grossfield and Shooter 1971) and rabbit (Cain, Ball, and Dekaban 1972). Similar techniques have shown differences among the regions of the auditory pathways in the guinea pig (Davies 1970) and between neuronal membrane fractions of DBA and C57 mice (Gurd, Mahler, and Moore 1972). In our laboratory, Caplan, Cheung, and Omenn (1974) have reported some differences in the electrophoretic profile of aqueous-soluble cortical human proteins between adults and fetuses or infants; in addition, in an effort to fractionate the protein to yield glycoproteins another difference was identified. Glycoproteins were isolated by affinity chromatography on columns of agarose to which the substance Concanavalin A was covalently bound. Concavalin A has high affinity for glycoproteins. A single band on polyacrylamide gels was obtained with the fraction eluted from the column with alpha-D-methyl-glucoside, while an additional band was obtained with a fraction eluted similarly but derived from fetal samples.

Such studies are frustrating in that differences which are identified cannot be interpreted in terms of functional pathways. The brain-specific proteins S-100 and 14-3-2 still lack functional assignments. However, some techniques are now available that may allow assignment of function to various protein bands. These techniques include reaction with specific antibodies against the S-100 protein of glial cells and the 14-3-2 protein of neurons; binding of radioactive ligands, such as colchicine for microtubular protein (Feit and Barondes 1970) and various small molecules for transport proteins; and incorporation in vivo of labeled amino acids (Packman, Blomstrand, and Hamberger 1971) and sugars (Zatz and Barondes 1970; Quarles, Everly, and Brady 1972) into metabolically active proteins and glycoproteins.

The potentially much more informative, but technically more demanding approach to proteins involves studies of enzymes and other proteins of known function. For example, the enzyme creatine phosphokinase (CPK) is known to be very important in generating high-energy ATP from stores of creatine-phosphate, especially in brain and muscle. Upon electrophoresis and specific staining for this enzyme, the CPK activity is localized in gels in readily distinguished positions when extracts of brain and of muscle are compared. A striking developmental transition in phenotype of CPK in muscle (but not brain) has been described (Eppenberger et al. 1964): Early fetal samples of skeletal muscle give the electrophoretic phenotype of normal adult or fetal brain, with a transition to the adult muscle pattern as gestation proceeds. The brain and muscle forms of the enzyme are termed *isozymes* and are almost surely the homologous products of two different genes produced by duplication of a precursor gene. The polypeptide chains of the isozymes have differences in their amino acid sequence, accounting for the difference in net charge that allows their separation upon electrophoresis. Many other sets of tissue-specific isozymes have been found for other enzymes of importance in metabolism. We have been particularly interested in the glycolytic enzymes, since glycolysis is so important to the basal metabolic integrity of the nervous system. Of the eleven enzymes in the glycolytic pathway, eight seem to have tissue-specific isozyme forms (see Omenn and Motulsky 1972). Also, it may be noted that the primary initial enzyme of the pathway is different in liver, that employs phosphoglucomutase to act on glucose-1-phosphate (from glycogen) as primary substrate, than in erythrocytes, muscle and brain, which employ hexokinase to act on glucose itself as primary substrate. In each case glucose-6-phosphate is produced for subsequent enzymatic conversion.

In our recent studies of glycolytic enzymes we have noted a particularly interesting isozyme phenomenon in the case of the enzyme phosphoglycerate mutase, PGAM (Omenn and Cheung 1974). This enzyme is somewhat analogous to CPK in that tissue-specific isozymes for brain and muscle can be readily distinguished and that the muscle in young fetuses has the phenotype of adult or fetal brain PGAM. We gave particular attention to the time course of transition of fetal to hybrid to adult muscle-type pattern of both PGAM and CPK in the same specimens; the pattern appears not to be synchronous. Just what physiological changes or genetic programme in skeletal muscle directs these transitions in gene expression during gestation are not known. A complementary phenotypic transition for PGAM has been noted in malignant transformation of brain tissues (Omenn and

Cheung 1974). In certain tumors, particularly the more malignant tumors, the electrophoretic pattern demonstrates the expression of muscle-type PGAM, as well as the normal brain-type enzyme. This kind of derepression or activation of an otherwise inactive gene may result in the "ectopic" appearance of polypeptides with hormonal, antigenic, or (in this case) enzymatic properties that permit their detection.

The model for developmental changes in a specific protein, of course, is the hemoglobin system in red blood cells and their precursors in the bone marrow (Motulsky 1969). A series of genes for the polypeptide chains of hemoglobin yield embryonic, fetal, and two kinds of adult hemoglobins. In muscle, as noted above, two excellent examples do exist, namely the CPK and PGAM systems. And in brain, the glycolytic enzyme aldolase is known to begin with a "common" phenotype found in muscle and liver and then to develop a brain-specific pattern due to expression of aldolase C. The aldolase system undergoes another transition in liver, with appearance of aldolase B. Muscle retains the aldolase A phenotype. The three aldolase systems have strikingly different substrate affinities, corresponding in the case of muscle to emphasis on glycolysis and in the case of liver to emphasis on gluconeogenesis. The properties of aldolase C are intermediate (Penhoet, Rajkumer, and Rutter 1966). More intensive search for additional examples of developmental transitions for functionally important emzymes in brain may be productive. The enzymes and transfer RNA's involved in protein synthesis and neurotransmitter-related enzymes may merit special attention. Since the enzyme dopamine beta-hydroxylase is an excellent marker for adrenergic fibers, we intend to determine the developmental time course in appearance of adrenergic fibers in cortex of our fetal specimens by assaying for the marker enzyme.

The same kinds of protein and enzyme analyses may be applied to the endocrine glands. Some important enzymes, such as the hydroxysteroid dehydrogenases, may be amenable to electrophoretic screening with specific enzyme stains. As far as I am aware, no data have been gathered on these tissues either for developmental transitions or for polymorphisms.

Molecular, Biochemical, Metabolic Differences: Normal Variation

It is very likely that the remarkable ranges of behaviors we observe in normal people have their counterpart at the metabolic and physiological level. It is not so clear to what extent observable metabolic variation represents the substrate for behavioral variation and to what extent metabolic variation represents the responses to learning and exogenous factors. In principle, the same kinds of parameters that were discussed above in the characterization of gene action in the brain might be examined across individuals to assess variation in both quantitative and qualitative properties. Quantitative variation is fraught with complications of measurement and known physiological and laboratory effects. Qualitative variation, particularly alteration of electrophoretic mobility and potentially drastic changes in catalytic or inhibitory activity, can be characterized much more definitively and is more likely to represent inherited alterations in the macromolecules themselves.

In the case of DNA we would expect to find no detectable differences between individuals with available techniques. The homologies simply swamp the expected variations in base sequences. In the case of RNA's there are greater opportunities to find at least quantitative differences, either in the proportions of the RNA's or the fractions of hybridizable RNA messengers, if competition experiments can be set up. Furthermore, electrophoretic profiles of RNA's and of ribosomal protein subunits are quite feasible. I am unaware of any attempts thus far to apply these techniques to the question of individual variation.

For enzymes, starch gel and acrylamide gel electrophoretic techniques have been highly developed for screening significant numbers of individuals for polymorphisms. Often the "isoalleles" of polymorphic systems have differing quantitative activity, indicating that these alleles may not be quite equivalent in the metabolic setting. For example, G6PD A+ has 85 percent and G6PD A− has 15 percent of the activity of the usual G6PD B form; the three alleles of acid phosphatase, occurring in six sets of dimer pairs, have relative activities of 100, 150, and 200. The acid phosphatase model is especially instructive, since a quantitative survey of human populations suggests a normal distribution of this enzyme activity in red blood cells; only with electrophoretic differentiation of the six dimeric phenotypes (AA, AB, AC, BB, BC, CC) can each subgroup be tested and be shown to have narrow ranges of enzyme activity (Harris 1970). Unfortunately, an in vivo role of this interesting enzyme has not been elucidated.

Study of polymorphic enzyme systems involving crucial metabolic processes in brain seems a potentially fruitful approach to the polygenic pattern of inheritance inferred from statistical analysis of behavioral traits. The electrophoretic screening method is capable of uncovering qualitative, structural differences in specific enzymes of different individuals, without confusion by different quantitative activity in different parts of the brain or upon physiological stimuli. However, the interpretation of the physiological consequences of these qualitative enzyme differences still will require careful measurement of the metabolic impact upon individuals having the two different types of enzymes. In humans such measurements must be carried out indirectly with radioactive tracers and with enzyme inhibitors; in model systems in mice or monkeys more direct measurements may be feasible.

The statistical notion that polygenic inheritance involves the equal and additive effects of a great many genes must be modified in light of metabolic interactions. Certain metabolic control points will be more important than others and much more important than enzyme reactions in minor pathways. Thus, it is possible that, even though a great many genes can interfere with normal brain development if completely deficient, the socalled normal range of development and function may be determined by a relatively few polymorphic genes sitting at rate-limiting steps in key metabolic pathways. The fact that a normal or Gaussian distribution of some quantitative variable is obtained does not require a large number of genes for explanation, as has been noted for the acid phosphatase system. The point of this discussion is to encourage the search for major gene mechanisms in polygenic traits and psychiatric disorders.

Thus far our attention has been directed primarily toward the energy-generating metabolic processes of the nervous system, recognizing the exquisite sensitivity of the brain to lack of glucose or oxygen. We have screened all eleven enzymes of the glycolytic pathway from hexokinase to lactate dehydrogenase in some 150 human brain specimens (Cohen et al. 1973). A polymorphism of a rate-limiting enzyme, such as phosphofructokinase in the glycolytic pathway, would be highly significant even if associated with only a small difference in quantitative enzyme activity, since production of lactate at the end of the pathway and of ATP along the way would be affected. On the other hand, a small difference in activity of an enzyme normally present in concentrations well above rate-limiting activities could be expected to have no such consequences. None of these enzymes has a common variant form. Only single, rare variants of phosphoglycerate kinase and of enolase were found, presumably reflecting mutations.

The above negative finding appears to be highly significant with regard to basic questions about genetic variation: It is likely that the very old evolutionary status of glycolysis and its central role as the primary pathway of glucose utilization in the brain have placed remarkable constraints on the tolerance for mutation-induced variation in the protein structure of these enzymes. Most of the glycolytic enzymes, as noted above, have evolved tissue-specific isoenzyme forms. From the point of view of identifying biochemical variation for correlation with behavioral traits, however, this particular study failed to identify any such variations. One might rationalize that the absence of variation in an essential pathway is reasonable and may make such variation as is found in certain other, perhaps evolutionarily less entrenched, pathways all the more useful.

Our screening did yield one highly polymorphic system, the malic enzyme or NADP-linked malate dehydrogenase (Cohen and Omenn 1972). Studies with human and monkey brain tissue demonstrated that the cytoplasmic and mitochondrial forms of the malic enzyme are controlled by different genes and vary and segregate independently. The mitochondrial malic enzyme in man has three electrophoretic phenotypes in starch gel electrophoresis, corresponding to gene frequencies of 0.7 and 0.3 for the two alleles. This enzyme may be involved in hydroxylation reactions, particularly in the adrenal cortex and perhaps in the brain. Its three phenotypic forms were isolated and characterized biochemically without any demonstrable functional differences in vitro (Cohen and Omenn 1972).

We are expanding the study of enzymes to those involved in neurotransmitter metabolism and biosynthesis, with the expectation that biochemical correlates of neural plasticity are more likely to be found in such pathways than in the basic energy-generating processes. The usual approach of electrophoresis and specific stain for enzyme activity is not readily applicable to these enzymes, either because their activities in extracts are too low or because their products do not couple to stainable compounds. However, a second powerful experimental tool of the biochemical geneticist can be marshalled for this effort: pharmacogenetic analysis (Omenn and Motulsky 1972). Specific inhibitors are known and available for almost every step in biosynthesis and metabolism of norepinephrine, acetylcholine, and gamma aminobutyric acid.

Our first set of experiments focused on monoamine oxidase, even though there is some evidence that isozymes exist in brain; certainly multiple molecular forms are detected in vitro. The assays of the enzyme in homogenates for susceptibility to inhibition by pargyline and by D-amphetamine failed to show any individuals with distinctly different inhibition constants for monoamine oxidase activity. Also, there was not a significant difference between adults and infants. Further work with isozymes may be indicated. We are also working at generating high enough activity of dopa decarboxylase, tyrosine hydroxylase, and dopamine beta-hydroxylase to follow their decrease in activity upon addition of specific inhibitors.

Another important source of potential variation is the hormonal environment associated with sex. In fact, one of the especially timely questions about human behavior is the issue of male/female differences and the extent to which they reflect cultural impact of the assigned sex role or biological impact of the sex chromosomes and sex hormones. As noted below, certain disorders of sexual differentiation allow some critical questions in this regard. In general, however, sex is an important variable to analyze in studies at all levels of the nervous system. Obviously the interaction of nervous system and endocrine functions must be analyzed separately by sex.

Finally, too often neglected in studies of the brain are the sense organs, for which powerful physical tools and some potential biochemical analyses are available. As progress has been made in mapping the detailed representation of visual stimuli and of input to the cerebellum, variation and defects in these pathways should be sought. Even such a well recognized, reliably measured, and common abnormality as color blindness has been rather little investigated from the behavioral side.

Mutant Analysis of Human Behavior

The inborn errors of metabolism must be divided, for purposes of understanding their mechanisms, into those acting extrinsically and those acting instrinsically to interfere with normal brain development. The case of phenylketonuria is really a toxic influence on the brain, not very different from lead poisoning. It should be noted that Weber (1969) postulated that postnatal brain was especially sensitive to phenylalanine and phenylketone levels because of effects of these compounds on hexokinase and pyruvate kinase having either lower activity or higher affinity than found in older persons. I do not know of any confirmation of that report. Dr. Elving Anderson has carried out a number of interesting psychological and perceptual-motor skill tests in children with phenylketonuria. Another interesting amino aciduria is histidinemia, in which about half of the children have been reported to have speech defects or impediments, often but not necessarily in association with mild or moderate mental retardation. Dr. Barton Childs has been particularly interested in this syndrome.

The disorders that are intrinsic to the nervous system bear special attention. We may note the Lesch-Nyhan syndrome, homocystinuria, and the adult form of metachromatic leukodystrophy. The Lesch-Nyhan syndrome is comprised of hyperuricemia, choreoathetotic movement disorder, and a self-destructive, impulsive behavior. It is due to deficiency of an enzyme known as hypoxanthine-guanine phosphoribosyltransferase (HGPRT), involved in what was previously discounted as a minor "salvage" pathway of purine metabolism. This enzyme turns out to have its highest activity in the body in the basal ganglia of the brain, allowing a correlation with the neurologic disorder of choreoathetosis. Just how to relate this metabolic disorder to an uncontrollable impulse to bite off the fingertips or the lips is a challenge to investigators of impulsive or violent behavior disorders in man. Attempts to lower the hyperuricemia itself from birth have not averted the development of the behavioral aberration (Nyhan 1972). Incidentally, the importance of this metabolic pathway is underscored by the finding that heterozygous females (cellular mosaics by the process of random X-inactivation for this X-linked trait) have the expected 50 percent of normal activity for HGPRT in skin fibroblasts, but 100 percent of normal activity in blood cells (Nyhan et al. 1970). Presumably, all blood cell precursors lacking HGPRT activity were eliminated. We have no information on HGPRT-negative cells in the nervous system.

A second remarkable metabolic error intrinsic to the nervous system, as well as other tissues, is homocystinuria, due to deficiency of the enzyme cystathionine synthetase. As a result, cystathionine is not formed and homocysteine and methionine accumulate. Cystathionine is a complex amino acid found normally in remarkably high concentrations in the brain, but its function is entirely unknown. A different inborn error, cystathioninuria, due to deficiency of the enzyme to break down cystathionine, seems to be unassociated with any major defects. Clinically, homocystinuria is characterized by vascular thromboses, skeletal anomalies, downward displacement of the ectopic lens of the eye, and—in only about one-half of cases—mild to moderate mental retardation. There is considerable dispute whether affected patients or their sibs might have an increased incidence of schizophrenia; the evidence is not impressive, but the speculation seemed sound, based upon the hypothesis that methylated derivatives of normal neurotransmitter substances might be pathogenetically involved in schizophrenia or at least in some experimental hallucinatory states. Why only one-half of cases have mental retardation is unclear. Perhaps the others have lower IQ than would have been their potential, but are still within the normal range. Perhaps the enzyme defect is different in different individuals. In fact, at least two types of biochemical defect have been recognized, since one group of these patients can be helped with massive doses of the co-factor pyridoxine or vitamin B6. Nevertheless, both the pyridoxine-responsive and pyridoxine-unresponsive patients may have mental retardation. Cerebral vascular accidents may be responsible.

In metachromatic leukodystrophy, severe neurological deficit and poor mental function is associated with complete enzyme deficiency and infantile onset. However, at least nineteen cases have now been reported with late adolescent or early adult onset and a sufficiently mild involvement of neurologic function that psychological difficulties dominate the clinical picture. In all of these cases, the patients were institutionalized with a clinical diagnosis of schizophrenia, only later to develop neurologic signs and be found to contain huge amounts of stored sulfatide in the brain. This disorder is important for two reasons: First, because it contributes slightly and serves as an example of many other possible entities that may make up the icebergs of the major psychiatric diagnoses; and second, because it illustrates that much milder symptoms or even variation within what we call the range of normal may be caused by or associated with mild deficiencies of those enzymes intrinsic to the brain for which severe deficiency causes gross interference with normal development. These rare autosomal recessive disorders with a frequency of, say, 1 in 40,000 births each would have a carrier frequency of 1 percent. For certain enzymes present in the brain and present in near rate-limiting activity, such a decrease to 50 percent of normal activity in the carrier might be a significant factor in mild mental impairment or polygenic traits or regionally-specific mental defects. There has been very little detailed psychometric study of such possibilities.

Dr. Anderson has, in fact, initiated a series of studies of manual dexterity and other specific functions in carriers as well as patients with the PKU gene (Anderson et al. 1968, 1969).

Application to other less common syndromes, possibly through collaborative efforts of several centers, might be rewarding. The only definite lead to regional defects comes not from a metabolic error, but from the chromosomal anomaly 45,XO or Turner's syndrome, in which Money (1963) demonstrated a striking defect in space-form perception and in drawing ability. The clinical counterpart for that psychometric anomaly is readily recounted by gynecologists who have had to direct such young women through the maze of partitions in the usual out-patient clinic! It would be worth checking whether some region of parietal cortex requires two active X chromosomes in females, as does the normal ovary.

Abnormalities of the endocrine systems also are known to affect behavior. The cases of full-blown cretinism or hypoparathyroidism are well recognized, but mild defects in adrenal, gonadal, parathyroid, pancreatic, and pituitary function have been little analyzed behaviorally. Testicular feminization, a syndrome in which genetic males with normal testes and normal production of testosterone fail to become masculinized because the target tissues fail to respond to the hormone, is now being studied in an animal model (Dofuku, Tettenborn, and Ohno 1971). These individuals act and appear as females. Conversely, females with the adrenogenital syndrome tend to become masculinized as a result of fetal exposure to circulating androgens (Money and Ehrhardt 1972). However, my colleague McGuire and I in Seattle have been unable to demonstrate that the adrenogenital syndrome was associated with any specific increase in IQ, as has long been claimed, when the expected IQ based upon parents and siblings was determined (also see Chapter 7).

In the development of behavior, timing and the pace of development must be as important as any consistent structural abnormality. So-called "immaturity" or "developmental lag" can have serious consequences comparable to the late appearance of the bilirubin-conjugating system in liver. Late appearance of language or prolonged difficulty in learning to read is often attributed to delayed maturation. Very carefully gathered, age-corrected data for neurophysiological responses and biochemical transitions may prove productive in this area. As in all behavioral syndromes, heterogeneity of mechanisms must be anticipated. Families or kindreds with several affected members may reward intensive study far better than a much larger number of unrelated affected individuals.

How Much Will Be Learned About the Brain From Biochemical Studies?

Some biologists, evolutionists, and philosophers view the nature of man and of his consciousness as a complexity beyond human understanding (Eccles 1967). There may be a kind of Maxwell's demon in all our studies, in that every type of probe perturbs the system. Most important, as discussed in the introduction to this chapter the cell-cell interactions, cell death, and even the biochemical expression of individual cells may reflect stochastic processes. We are probably far short of having identified all the chemical neurotransmitters; yet, recently electrically-mediated synapses have been demonstrated. In cell culture, mouse neuroblastoma cells clearly have the potential to express either adrenergic or cholinergic properties, both enzymatic and neurotransmitter properties. Although it is not altogether clear, an individual cell derived from such a clone seems to produce one or the other transmitter, not both. The factors that determine this difference are not elucidated, though cell density, cyclic AMP, and other media factors can turn up or turn down neurotransmitter-related enzyme activity for the system present in the cells (Amano, Richelson, and Nirenberg 1972).

While individual investigators usually must focus on a particular level of inquiry—anatomical, metabolic, molecular, neurophysiological, psychometric, clinical—it is important that many of these disciplines be combined in well organized interdisciplinary studies of the precious specimens and patients potentially available for study. For the next decade, I am convinced that much more will be learned about brain function from biochemical and pharmacological studies and neurophysiological monitoring than from all the statistical analyses of behavioral measures that could be imagined. Even though total understanding may not be possible, the potential to increase our knowledge of human behavior by detailed analyses and by integrated and comparative study of complex behaviors offers excitement and challenge for experimental exploration of the development and function of the nervous system.

References

Amano, T.; Richelson, E.; and Nirenberg, M. 1972. Neurotransmitter synthesis by neuroblastoma clones. *Proceedings of the National Academy of Science* 69: 258-63.

Anderson, V. E.; Siegel, F. S.; Tellegen, A.; and Fisch, R.O. 1968. Manual dexterity in phenylketonuric children. *Perceptual and Motor Skills* 26: 827-34.

Anderson, V.E.; Siegel, F.S.; Fisch, R.O.; and Wirt, R.D. 1969. Responses of phenylketonuric children on a continuous performance test. *Journal of Abnormal Psychology* 74: 358-62.

Cain, D.F.; Ball, E.D.; and Dekaban, A.S. 1972. Brain proteins: Qualitative and quantitative changes, synthesis and degradation during fetal development of the rabbit. *Journal of Neurochemistry* 19: 2031-42.

Caplan, R.; Cheung, S.C-Y; and Omenn, G.S. 1974. Electrophoretic profiles of aqueous-soluble proteins of human cerebral cortex: Population and developmental characteristics. *Journal of Neurochemistry* 22: 517-20.

Cicero, T.J.; Cowan, W.M.; Moore, B.W.; and Suntzeff, V. 1970. The cellular localization of the two brain specific proteins, S-100 and 14-3-2. *Brain Research* 18: 25-34.

Cohen, P.T.W.; and Omenn, G.S. 1972. Variation in cytoplasmic malic enzyme and polymorphism of mitochondrial malic enzyme in Macaca nemistrina and in man. *Biochemical Genetics* 7: 289-301, 303-11.

Cohen, P.T.W.; Omenn, G.S.; Motulsky, A.G.; Chen, S-Y; and Giblett, E.R. 1973. Restricted variation in the glycolytic enzymes of human brain and erythrocytes. *Nature New Biology* 241: 229-33.

Davies, W.E. 1970. The incorporation of (14C) into the protein of the guinea pig central auditory system. *Journal of Neurochemistry* 17: 297-303.

Dofuku, R.; Tettenborn, U.; and Ohno, S. 1971. Testosterone —"regulons" in the mouse kidney. *Nature New Biology* 232: 5-7.

Eccles, J.C. 1967. Evolution and the conscious self. In *The Human Mind. Nobel Symposium*, ed. J.D. Roslansky. Amsterdam: North Holland Publication, pp. 1-28.

Eppenberger, H.M.; Eppenberger, M.; Richterick, R.; and Aebi, H. 1964. The ontogeny of creatine kinase isoenzymes. *Developmental Biology* 10: 1-16.

Feit, H., and Barondes, S.H. 1970. Colchicine-binding activity in particulate fractions of mouse-brain. *Journal of Neurochemistry* 17: 1355-64.

Grossfield, R.M., and Shooter, E.M. 1971. A study of the changes in protein composition of mouse brain during ontogenetic development. *Journal of Neurochemistry* 18: 2265-77.

Grouse, L.; Chilton, M.D.; and McCarthy, B.J. 1972. Hybridization of ribonucleic acid with unique sequences of mouse deoxy ribonucleic acid. *Biochemistry* 11: 798-805.

Grouse, L.; Omenn, G.S.; and McCarthy, B.J. 1973. Study by DNA/RNA hybridization of the trancriptional diversity of human brain. *Journal of Neurochemistry* 20: 1063-73.

Gurd, R.D.; Mahler, H.R.; and Moore, W.J. 1972. Differences in protein patterns on polyacrylamide gel electrophoresis of neuronal membranes from mice of different strains. *Journal of Neurochemistry* 19: 553-56.

Hahn, W.E., and Laird, C.D. 1971. Transcription of nonrepeated DNA in mouse brain. *Science* 173: 158-61.

Harris, H. 1970. *Principles of human biochemical genetics*. Amsterdam: Elsevier.

McCarthy, B.J., and Hoyer, B.H. 1964. Identity of DNA and diversity of messenger RNA molecules in normal mouse tissues. *Proceedings of the National Academy of Science U.S.* 52: 915-22.

Money, J. 1963. Cytogenetic and psychosexual incongruities with a note on space-form blindness. *American Journal of Psychiatry* 119: 820-27.

Money, J., and Ehrhardt, A.A. 1972. *Man and woman/boy and girl*. Baltimore: Williams and Wilkins.

Moore, B.W.; Perez, V.J.; and Gehring, M. 1968. Wallerian degeneration in rabbit tibial nerve: Changes in amounts of the S-100 protein. *Journal of Neurochemistry* 15: 971-77.

Motulsky, A.G. 1969. Biochemical genetics of hemoglobins and enzymes as a model for birth defects research. *Proceedings of the 3rd International Congress of Congenital Malformations*, pp. 199-208. The Hague: Excepta Medica.

Nyhan, W.L. 1972. Disorders of nucleic acid metabolism. In *Biology of brain dysfunction*, ed. G. Gaull, vol. 1, pp. 265-300. New York: Plenum Press.

Nyhan, W.L.; Bakay, B.; Connor, J.D.; Marks, J.F.; and Keele, D.K. 1970. Hemizygous expression of glucose 6-phosphate dehydrogenase in erythrocytes of heterozygotes for the Lesch-Nyhan syndrome. *Proceedings of the National Academy of Science, U.S.* 65: 214-18.

Omenn, G.S., and Cheung, S. C-Y. 1974. Phosphoglycerate mutase isozyme marker for tissue differentiation in man. *American Journal of Human Genetics* 26: 393-99.

Omenn, G.S., and Motulsky, A.G. 1972. Biochemical genetics and the evolution of human behavior. In *Genetics, Environment and Behavior*, eds. L. Ehrman, G.S. Omenn, and E. Caspari, pp. 129-79.

Packman, P.M.; Blomstrand, C.; and Hamberger, A. 1971. Disc electrophoretic separation of proteins in neuronal, glial and sub-cellular fractions from cerebral cortex. *Journal of Neurochemistry* 18: 479-87.

Penhoet, E.; Rajkumer, T.; and Rutter, W.T. 1966. Multiple forms of FDP aldolase in mammalian tissue. *Proceedings of the National Academy of Science, U.S.* 56: 1275-82.

Quarles, R.H.; Everly, J.L.; and Brady, R.O. 1972. Demonstration of a glyco-protein which is associated with a purified myelin fraction from rat brain. *Biochemical and Biophysical Research Commun.* 47: 491-97.

Shooter, E.M. 1972. Some aspects of gene expression in the nervous system. In *The Neurosciences, 2nd Study Program*, ed. F.O. Schmitt, pp. 812-26.

Sotelo, C., and Paley, S.L. 1968. The fine structure of the lateral vestibular nucleus in the rat, I. Neuron and neurologlial cells. *Journal of Cell Biology* 36: 151-79.

Weber, G. 1969. Inhibition of human brain pyruvate kinase and hexokinase by phenylalanine and phenylketonuric brain damage. *Proceedings of the National Academy of Science, U.S.* 63: 1365-69.

Zatz, M., and Barondes, S.H. 1970. Incorporation of mannose into mouse brain lipid. *Journal of Neurochemistry* 17: 157-63.

Commentary I

V. Elving Anderson
University of Minnesota

Dr. Omenn has summarized in this chapter a wide range of detailed findings about nervous system functioning; there are some important general implications that should be clear to the reader:

1. A knowledge of biochemistry is becoming essential for students of behavior.
2. There are many different mechanisms by which genetic factors may influence the development and function of the nervous system. The belief that genes exert no influence upon behavior has become completely untenable.
3. Even though all nucleated cells in an organism contain the same genetic material, not all of the genes are expressed. We can begin to understand how genes are turned off or on at different stages of development and in different tissues.
4. The abundant evidence for individual variability at the genetic level can strengthen other lines of evidence for individual differences in behavior.
5. An appreciation of genetic variability will help investigators to plan more adequate ways to study the influence of environmental factors upon behavior.

I would like to comment in more detail on some of Dr. Omenn's ideas. One is the concept of the "localization of brain function" in specific regions. The brain is no longer a "black box," but is known to have a complex internal structure. A number of separate neuronal systems have been identified in the rat brain, some using dopamine as a neurotransmitter and others, using serotonin (Ungerstedt 1971b). Even within cells a compartmentalization of structure and function can be elucidated, as Dr. Omenn mentioned.

At the 1974 meetings of the American Society for Neurochemistry there have been several interesting papers concerning brain localization with respect to tyrosine hydroxylase. The activity of the enzyme was increased by the *addition* of calcium ions to medulla-pons homogenates, but was also increased by the *removal* of calcium from homogenates of corpus striatum (Morgenroth, Boadle-Biber, and Roth 1974). The caudate nucleus had a low molecular weight form of the enzyme, the locus coeruleus contained a high molecular weight form, while the substantia nigra had both types (Joh, Pickel, and Reis 1974).

Another illustration comes from the psychotic disorders. C.D. Marsden (1973) pointed out that brain amines are studied by neurologists to account for abnormal movements and by psychiatrists to explain the major psychoses. He went on to suggest that the basal ganglia may be just as concerned with emotion, thought, and intellect, as with motor control. Matthysse (1973) reviewed various hypotheses to account for the action of phenothiazines, comparing those drugs having antipsychotic effects (chlorpromazine and haloperidol) with those having no effect on psychotic manifestations (promethazine). He then applied this type of reasoning to evaluate the various brain areas that have been implicated in the etiology of schizophrenia. In our efforts to understand possible genetic mechanisms that might lead to such behavioral pathology we must consider changes affecting particular neurotransmitters, pathways, or brain areas.

In the Lesch-Nyhan syndrome, mentioned by Dr. Omenn, there are two independent lines of evidence that suggest that the basal ganglia may be involved: (1) The enzyme involved in the syndrome has an activity level in the basal ganglia (of normal brains) that is two to three times that found in other brain areas; (2) compulsive gnawing behavior in rodents is associated with lesions of the basal ganglia. Such behavior was induced in rats when crystalline DOPA or apomorphine was implanted in the dorsal part of the caudate nucleus or in the globus pallidus (Ernst and Smilek 1966). However, no response was observed with implants in the ventral part of the caudate nucleus, in the substantia nigra, or in other parts of the brain.

In another study, Ungerstedt (1971a) observed rats in which the nigro-striatal dopamine pathway had been destroyed on one side of the brain. Unilateral self-gnawing was induced by apomorphine, but was prevented by prior application of haloperidol. High doses of methylphenidate stimulated stereotypy (including continuous biting or gnawing), but this effect was abolished by lesions of the globus pallidus (Costall and Naylor 1974). These observations lead to the hypothesis that the enzyme deficiency in the Lesch-Nyhan syndrome produces some type of damage in specific brain pathways. Biochemical correction of the enzyme deficiency may not be able to reverse the damage itself, but drugs that alter the sensitivity of the damaged area to neurotransmitter substances might reduce the self-mutilating behavior.

I would also like to emphasize the evidence from biochemical genetic studies for *development*, specifically for the differential expression of different parts of the genome over time and in different tissues. Dr. Omenn's illustrations were selected primarily from studies at the molecular level, but a related set of examples can be chosen for genetic control of the structural development of the nervous system. Detailed information is now available for some of the neurological mutants in the mouse, primarily those involving the cerebellum. More recently, the analysis of induced mutations in *Caenorhabditis elegans* (a small roundworm) has produced direct evidence for genetic control of the "wiring diagram" of the nervous system (Brenner 1973, 1974).

Another interesting observation is that the gene for albinism affects nervous system development. The proportion of the nerve fibers from the retina of one eye that cross over to the other side of the brain is about 90 percent for rodents and close to 50 percent for humans. In albinos of eight mammalian species (including Siamese cats and tigers) the number of fibers passing from an eye to the ipsilateral cerebral hemisphere is reduced and the fiber endings are disorganized (Kaas and Guillery 1973). Studies in human albinos using cortical evoked potentials strongly suggest a disturbance of fiber pathways as compared with normals (Witkop 1974).

In some cases current views about the mechanisms of gene action may need reexamination. There are about thirty "shaker-waltzer type" mutations in the mouse that lead to a modification of behavior including head-jerking, hyperactivity, and circling. Some mutant animals show deafness or changes in the inner ear, and it has been assumed that these sensory defects are a sufficient cause for the behavioral anomalies. Deol (1966), however, studied mice of several shaker-waltzer type mutant strains and found that no abnormality of behavior can be ascribed to any particular defect or combination of defects of the inner ear. On the basis of this evidence, he suggested that the behavioral anomalies are more likely to arise from the genetic effects upon the central nervous system. More recently, Cools (1972) has found that some aspects of the behavior in varitint-waddler mice resemble the behavior induced by amphetamine in nonmutant mice and that ^{14}C-labelled DOPA is attracted preferentially to the Harderian gland of waltzing mutants (but not for nonwaltzing mutants or nonmutants). Cools concluded that a hyperactive dopaminergic mechanism may be involved. It is possible that the shaker-waltzer type mutants may provide valuable model systems for the analysis of sterotyped behavior (such as that observed in Parkinsonism or some psychotic disorders).

In passing, one point missing in Dr. Omenn's discussion is the recognition that a total life-span view of development should include aging. It is true that there are relatively few studies of genetic factors affecting the aging process, but the topic should be identified for future emphasis.

Next, I would like to comment on Dr. Omenn's discussion about "polymorphism." Individual variability is viewed by some investigators as noise in the system, to be treated statistically as error variance and to be reduced by pooling data from many subjects. This is not the way a geneticist approaches variability, as Dr. McClearn mentioned earlier (see Chapter 1). For this reason Dr. Omenn's analysis of brain enzymes is very informative. He found essentially no variability in those enzymes that are involved in the basic energy pathways, but did find variation in other enzyme systems.

These results provide an excellent illustration of the two general principles stated by Johnson (1974) in his review of available data from a number of species: (1) Enzyme polymorphism may parallel the variation in substrates that can be used. Enzymes of broad specificity (many of which act upon substrates originating from the external environment and thus potentially variable from time to time and from place to place) are far more variable than those enzymes that utilize specific metabolically-produced substrates (such as glucose); (2) enzyme polymorphism is often associated with regulatory reactions. In a biochemical pathway the primary control usually is exercised at one or a few key reactions. Individuals who are heterozygous at loci for such enzymes have two different forms of the enzyme available for use. Thus, they may be more able than others to modulate the key reaction(s) as a means of compensating for variable external conditions. Both of these principles would appear to emphasize the potential importance of enzyme polymorphism for developmental studies of behavior.

Dr. Omenn stressed the value of a "mutant analysis" of behavior. In biochemical genetics we find that we can understand normal pathways by studying alterations in mutant organisms. In those forms of behavior that can be assessed quantitatively but for which genetic control has not yet been established, it would be desirable to consider routinely individuals at either extreme of the behavioral continuum as potential probands for family studies. A further step would be to select those families with two or more affected siblings as a population sample for more extensive biochemical or behavioral evaluation and then to analyze the data for variation within and between families (Anderson 1972).

When the mode of inheritance is known (if it turns out to be simple) the method of study becomes more straightforward. We have focused on phenylketonuria (PKU) as a possible model system for studying behavior genetics. (Anderson and Siegel, 1968; Anderson et al. 1969). Some of the effects are developmental and essentially irreversible (such as general level of IQ), while some behavioral features can be reversed by dietary modification even later in childhood. Efforts to alter the diet under controlled conditions have indicated individual variability, with the possibility that there may be three patterns of response to increased levels of phenylalanine (Frankenburg, Goldstein, and Olson 1973; Siegel, Anderson, and Bruhl 1967): (1) A continuing disturbance of behavior; (2) a transient effect, with a return to earlier, more normal, behavior within about two weeks; and (3) no noticeable effect. Possible explanations for the individual differences include the following: (1) Those children who are most damaged already may be more seriously affected; (2) there may be different mutations at the major genetic locus; (3) individuals may vary genetically in related enzyme systems that function sufficiently well in nonaffected children, but that are taxed by high levels of certain metabolites in those with PKU; (4) there probably are genetic differences in the nervous system response to the metabolic alterations, paralleling the known genetic differences in response to certain drugs. (I consider such studies as a part of "psychopharmacogenetics," thus adding a second connotation to the use of the term by Dr. Omenn.)

Finally, Dr. Omenn discussed the problems that arise when biochemical and behavioral variability are studied separately. One has either biochemical variability without functional correlations or behavioral variability without biochemical explanations. There is no easy answer, but there will be a premium on identifying and exploiting "model systems" with which one can study behavior, biochemistry, and genetics simultaneously.

Dr. Omenn's enthusiasm for approaches of this type is indicated by his claim that "much more will be learned about brain function from biochemical and pharmacological studies and neurophysiological monitoring than from all the statistical analyses of behavioral measures that can be imagined." Such a statement may appear controversial and does indeed warrant discussion as we seek to design and evaluate research studies. I suspect that he is right when the emphasis is on "brain function," but there are other questions about human behavioral development that deserve attention even though they cannot be approached directly in this manner at present.

References

Anderson, V.E. 1972. Genetic hypotheses in schizophrenia. In *Genetic factors in schizophrenia*, ed. A.R. Kaplan, pp. 490-94. Springfield, Ill.: Charles C. Thomas.

Anderson, V.E., and Siegel, F. 1968. Studies of behavior in genetically defined syndromes in man. In *Progress in human behavior genetics*, ed. S.G. Vandenberg, pp. 7-17. Baltimore: Johns Hopkins Press.

Anderson, V.E.; Siegel, F.S.; Fisch, R.O.; and Wirt, R.D. 1969. Responses of phenylketonuric children on a continuous performance test. *Journal of Abnormal Psychology* 74: 358-62.

Brenner, S. 1973. The genetics of behaviour. *British Medical Bulletin* 29: 269-71.

Brenner, S. 1974. The genetics of Caenorhabditis elegans. *Genetics* 77: 71-94.

Cools, A.R. 1972. Neurochemical correlates of the waltzing-shaker syndrome in the varitint-waddler mouse. *Psychopharmacology* 24: 384-96.

Costall, B., and Naylor, R.J. 1974. The involvement of dopaminergic systems with the stereotyped behaviour patterns induced by methylphenidate. *Journal of Pharmaceutical Pharmacology* 26: 30-33.

Deol, M.S. 1966. The probable mode of gene action in the circling mutants of the mouse. *Genetics Research, Cambridge* 7: 363-71.

Ernst, A.M., and Smilek, P.G. 1966. Site of action of dopamine and apomorphine on compulsive gnawing behaviour in rats. *Experientia* 22: 837-38.

Frankenburg, W.K.; Goldstein, A.D.; and Olson, C.O. 1973. Behavioral consequences of increased phenylalanine intake by phenylketonuric children: A pilot study describing a methodology. *American Journal of Mental Deficiency* 77: 524-32.

Joh, T.H.; Pickel, V.M.; and Reis, D.J. 1974. Localization of different forms of tyrosine hydroxylase to noradrenergic and dopaminergic neurons in brain. Paper presented at the American Society for Neurochemistry, 1974, New Orleans.

Johnson, G.B. 1974. Enzyme polymorphism and metabolism. *Science* 184: 28-37.

Kaas, J.H., and Guillery, R.W. 1973. The transfer of abnormal visual field representations from the dorsal lateral geniculate nucleus to the visual cortex in Siamese cats. *Brain Research* 59: 61-95.

Marsden, C.D. 1973. Neurology and psychiatry. *Psychological Medicine* 3: 265-66.

Matthysse, S. 1973. Antipsychotic drug actions: A clue to the neuropathology of schizophrenia? *Federation Proceedings* 32: 200-205.

Morgenroth, V.H. III; Boadle-Biber, M.C.; and Roth, R.H. 1974. Allosteric activation of striatal tyrosine hydroxylase by a calcium chelator. Paper presented at the American Society for Neurochemistry, 1974, New Orleans.

Siegel, F.S.; Anderson, V.E.; and Bruhl, H.H. 1967. The effect of diet change on position discrimination and reversals in phenylketonuria. *University of Minnesota Medical Bulletin* 38: 217-18.

Ungerstedt, U. 1971a. Stereotaxic mapping of the monoamine pathways in the rat brain. *Acta Physiologica Scandinavia 367,* suppl.: 1-48.

———. 1971b. Postsynaptic supersensitivity after 6-hydroxydopamine induced degeneration of the nigro-striatal dopamine system. *Acta Physiologica Scandinavia* 367, suppl.: 69-93.

Witkop, C.J. 1974. Personal communication.

Commentary II

*Barton Childs
The Johns Hopkins
University*

Dr. Omenn's discussion in this chapter is excellent. He has made the point that however behavior genetics is to be studied, whether by the usual approach of determining the average degree of inheritance of a particular phenotype, or by a search for the genetic origin of variations among individuals with defined phenotypes, or by the study of phenotypic properties of known genotypes, the ultimate answers must be biochemical. That is not to say, however, that we will not be required to accept descriptions at the clinical and physiological levels perhaps for a long time to come.

Studies of Gene Action in Brain

Dr. Omenn has described the complexity of the problem of tying particular behavioral characteristics to a particular gene or genes. He has made a good start with his studies of DNA-RNA hybridization, which, as he points out, suggests a wider repertory of metabolic functions in brain cells than in the cells of other organs. Intuitively this seems entirely reasonable, but other hypotheses should be tested. For example, is it possible that the extra RNA could represent multiple copies of genes; that is, duplicated loci capable of producing messenger RNAs?

Questions aside, this seems to me to be a signal advance and we will all be waiting to see whether the various parts of the brain, or the two sides, reveal different specificities.

Biochemical Differences in Normal Variation

Dr. Omenn has entered a fermenting field in looking at the extent of polymorphism in brain tissue. He has made the appropriate warnings about the difficulties of assigning behavioral attributes to particular polymorphic loci; but since the genes occuping these loci represent the material of which common variation must be made, they must underlie, or be related somehow, to common behavioral differences. Electrophoresis reveals that about one third of human gene loci are polymorphic and that each individual is heterozygous at 6 or 7 percent of his loci for these common genes (Harris and Hopkinson 1972). Since electrophoresis underestimates differences by perhaps a factor of three and since in addition each of us has some indefinable number of loci containing rare mutants, there is definitely plenty of genetic variation to conjure with.

The idea of individual differences in response to drug action was also mentioned, and Dr. Omenn should be encouraged to press on in this vein. Some such genetic variants are already known. For example, rapid or slow acetylation is associated with common alleles at the locus determining acetyl transferase in the liver, and these differences have been shown to be related to the therapeutic response to phenelzine in the treatment of depression (Evans and White 1964; Johnstone and Marsh 1973). In addition, familial responses have been shown to monoamine oxidase inhibitors and tricyclic drugs also in the treatment of depression (Pare 1970). All of this looks very promising, but its importance is only partly related to the prescription of particular drugs for particular people. Its more important implication is in the recognition of genetic variations in the metabolic disposition of the natural metabolites commonly encountered in cellular activity. After all these so-called detoxification mechanisms are called upon only occasionally to deal with exogenously administered drugs. Much more commonly, they are involved in the metabolism of normal substances. For example, some metabolites are conjugated with glucuronic acid, or with glycine, while others are acetylated, or sulfated, prior to excretion. Since the activity of these molecules may be changed by conjugation, variation in the rates of conjugation may be reflected in differences in the functions they govern, and if those functions influence behavior, the study of such properties would be a key to variations in normal behavioral phenotypes.

Mutant Analysis

This is the phenotypic characterization of known genotypes. Dr. Omenn mentions several diseases and there are, of course, many others (also see chapter 7). It is not surprising that diseases involving enzyme deficiencies in brain cells produce marked alteration in brain function, and patients who do not die of these disorders may be left markedly mentally retarded. Treatments for such diseases are being devised and where successful allow the development of what seems grossly to be normal intellectual powers. It will be interesting, however, to see whether all of the intellectual functions develop equally well or whether some such patients will show learning disabilities or other minor manifestations of disturbance in intellectual development.

These diseases are not likely, however, to be an important source of causes of learning disability since they are rare. On the other hand, even though they appear in the population with frequencies of 1 in 15,000 to 1 in 500,000, the possessors of these same genes in the heterozygous state are numerous and might provide a more promising hunting ground for genes that influence behavior.

We are presently witnessing an exponential increase in the description of inborn errors of metabolism and this proliferation will continue. Of something in excess of 120 inborn errors of metabolism whose enzyme deficiencies have been characterized, perhaps one fourth to one third cause mental retardation. Persons heterozygous for the genes causing these diseases might be worth looking at from the viewpoint of behavioral variability, especially those in which the enzyme deficiency is present in the neurons themselves. When one measures the activity in the cells of heterozygotes for enzymes which in the homozygous state show essentially no activity, one finds in the heterozygotes on the average about 50 percent activity. But this is an average and represents a distribution running from 20 or 30 percent in some persons to 60 or 70 percent in others. In some individuals enzyme activity of only 20 or 30 percent might make such steps rate-limiting when under normal conditions they are not, and this might be reflected in relatively minor but important variations in development and behavior.

If among the list of inborn errors there are thirty or forty diseases causing mental retardation in homozygotes, and if these disorders have an average frequency of about 1 per 100,000, then 0.5 to 1.0 percent of the population would be heterozygous for each one, which means that one fifth or more of the population would be heterozygous for one or other of these deficiencies. No one supposes that thirty or forty such disorders represent the total potential list, so these genes may be an important source of human variability. That this may be so is suggested by a recent study of parents of children with phenylketonuria (Ford and Berman 1974). Since the disease is a recessive both parents must be heterozygous for the gene contributing to a deficiency of phenylalanine hydroxylase. Standard IQ tests were done on 115 such parents and all were found to fall within the limits of normal variability. On the other hand, a highly significant inverse correlation was found between the IQ value and the height of the blood phenylalanine under the conditions of a standard tolerance test. That is, the more intolerant the parent was of phenylalanine, the lower was his IQ, even though the latter fell within the limits of normal. No studies were done to correlate verbal or other cognitive properties with the phenylalanine level, so we do not know whether the presumed effect of the gene was in any way selective.

This is a promising beginning that ought to be a prototype for many others. The way to proceed is simply to look at the behavioral attributes in standard psychometric contexts of persons possessing particular genes. Dr. Money has been doing this for some time with people who have chromosomal abnormality, but too few studies have been done of persons possessing particular genes.

References

Evans, D.A.P., and White, T.A. 1964. Human acetylation polymorphism. *Journal of Laboratory and Clinical Medicine* 63: 394-403.

Ford, R.C., and Berman, J.L. 1974. Intelligence and response to phenylalanine tolerance tests among parent carriers of phenylketonuric and hyperphenylalaninemic children. *Pediatric Research* 8: 389.

Harris, H., and Hopkinson, D.A. 1972. Average heterozygosity per locus in man; an estimate based on the incidence of enzyme polymorphisms. *Annals of Human Genetics* 36: 9-20.

Johnstone, E.C., and Marsh, W. 1973. Acetylator status and response to phenelzine in depressed patients. *Lancet* 1: 567-70.

Pare, C.M.B. 1970. Differentiation of two genetically specific types of depression by the response to antidepressant drugs. *Human Genetics* 9: 199-201.

Quantitative Genetic Perspectives: Implications for Human Development

L. L. Cavalli-Sforza
Stanford University

Our knowledge of behavioral genetics is still very limited but sufficient to convince us that behavior, like every other activity of men and animals, is at least to some extent under control of the genetic background. This does not, of course, exclude control from other variables, nor should it engender pessimism about the power of education, which I think would be a serious mistake in the light of all our present knowledge. In this chapter I want to examine some of the possibilities for research, and the major difficulties that are encountered both at the scientific and at the moral level.

The Moral Problem

It may be appropriate to start with the moral problem. The increasing rate of technological development, which is both a consequence of and a stimulus for scientific development, has made us acutely aware that every discovery in the laboratory has deep effects at some stage or other on life of the community. Behavior is almost synonymous with social life. This is especially true of an animal like man, even more so at a time when the rate of population growth is very high. With the increase of population densities, social interactions are destined to become more and more complex. Naturally, social interaction has been important for most of human evolution, but the increase in numbers and in communication that we are experiencing can only make it an even more pressing problem for the time to come.

Perhaps the simplest way to introduce the moral problem is to look at it from my personal point of view as a scientist engaged in the study of behavior. I find no moral worries about my own actions when I engage in research on abnormal behavior. Even if the threshold separating normal from abnormal is always a very difficult one to set exactly, there are extreme situations of harmfulness for society or to oneself that make a given type of behavior definitely undesirable. I can see no harm in any research that can help in understanding this kind of behavior, both at the

---

Preparation of this chapter was supported in part by Grant # NS 10711-02 from the National Institutes of Health.

genetic and at the environmental level. In fact, I believe it is one of the fields of research in which advance is most urgently needed. But the transition between abnormal and normal is quite gradual, and it may even be argued that the best way to understand normal behavior is by studying the extreme cases that are labelled as abnormal. It is therefore necessary to admit that research on abnormal behavior will have a bearing on normal behavior and vice versa.

When we consider the study of normal behavior, which in man is still a very difficult subject but is certainly making advances in animals, the moral problem is not so easy to answer anymore. If we try to anticipate where our research might lead us, we find two possible extremes. One of these poles is represented by the extreme hereditarian view; that everything we do is really determined by our genes and that there is nothing we can do about it. This type of genetic fatalism is rather distressing. Still, it is a possibility we may have to face because we do not yet know enough to predict what the main determinants of behavior are in the individual. The opposite pole is represented by the belief that we are so strongly subject to environmental influences that our individual genetic differences are trivial with respect to the power of the environment in which we develop. I find this belief equally worrying, because if true it means we are almost powerless in the hands of dictators or political conspirators who try to make us their slaves. *Brave New World* or *1984* are well known fictional examples that come to mind. Nevertheless, I have faith, which I hope will not turn out to be undue optimism, that we will avoid both these extremes, namely of genetic fatalism on the one hand and of behavioral control on the other. Perhaps one can derive some optimism from consideration of PKU (phenylketonuria), a perfect example of how knowledge of biochemical genetics can help prevent the development of a serious behavioral deficit once it is identified in time. This shows how, at least in principle, the danger of genetic fatalism can be eliminated. My belief that we will not fall into the trap of the opposite extreme is based on somewhat weaker grounds. It comes to a large extent from the impression that genetic variation in behavior is sufficient within our species to make it difficult to establish a complete control destined to last for a long time over a large society. A full discussion would lead me into areas that would probably take us away from the main issue.

I therefore find no moral bar to research in behavioral genetics, but I do believe cautions are necessary here more than elsewhere in order to avoid harm to individuals or groups. For this reason, I strongly disagree with the position of those who choose, on the basis of very weak evidence, to make sweeping statements on issues such as differences between groups for IQ and other traits (recall the recent controversy on IQ and race) that cannot fail to have important social consequences. That scientists have social responsibilities and should develop the necessary sensitivity to them is even more obvious in behavioral genetics than in other areas of research. There is a similar situation in statistical testing that may be worth noting. If I want to reject a null hypothesis and know that if I am wrong the cost is going to be very high, I will set a more stringent probability level for my decision. Similarly, one requires strong rather than weak evidence for making statements that might have sweeping social effects.

I felt I had to go into some detail on the moral aspects of research on behavioral genetics for various reasons. Recently, scientists have become more aware of the moral problems of their work and of its relevance for society. I think this is very desirable. On the other hand, I am also aware that it is very easy to be misunderstood. I had an example when the editor of the authoritative British journal, *Nature* (1970) accused Walter Bodmer and myself of being obscurantists because of an article that we published in *Scientific American* in the October 1970 issue on problems of intelligence and race. Our point was that there was more to be gained by looking at environmental factors determining IQ race differences than at genetic ones, given that the former are easier to study and more accessible to society's control if this is deemed desirable. This was taken to mean that we were for vetoing all sorts of behavioral genetics (see also our answer, Bodmer and Cavalli-Sforza 1971a). Proof that I am not an obscurantist as originally held by *Nature's* editor is, I hope, given by the fact that I am myself working in behavioral genetics. I am very grateful to have this opportunity of stating my views on some general objectives for this kind of research and its possible outcomes in the not too distant future, and also of reemphasizing that in behavioral genetics, more than in any subject, scientists should be aware of their limits and responsibilities.

Development, Genetics, and Environment

I should also offer an apology for using the word "development" in the title. Behavior is a problem of development more than perhaps any other one, but unfortunately there is a real paucity of information to review or of possibilities to consider. Man's behavior changes or develops progressively throughout his lifetime. The book called the *Seven Ages of Man* (edited by Sears and Feldman 1964) provides a summary of this developmental process. But it is in considering this type of problem that we feel most acutely the difficulties due to our short life span as investigators in comparison with the length of time that is needed for a proper longitudinal analysis. Maybe part of this concern is generated by my being in the sixth age of my life and therefore less likely to contribute anything by longitudinal studies. People younger than I may be bolder in this respect.

Some longitudinal material is now becoming available and it is of great interest. There is a divergence between the results of Wilson and of McCall on IQ development in twins. (McCall 1972; Wilson 1972). The results of the latter author are limited to very early ages (first and second years) at which tests do not necessarily measure the same type of intellectual performance that is measured by IQ tests at a later age. The analysis of longitudinal changes from two to eighteen years in the study by McCall (1970) shows us that this supposedly innate and supposedly well measurable quantity indicating our intellectual performance may change quite radically during our life. Some environmental influences have been detected in that study.

Environmental influences are naturally more interesting than the genetic ones from a social point of view when it comes to action. We may put in a first category those that have a chance of affecting more or less everyone in a similar way, such as in general nutrition or education. Some of these are obvious; in other cases however, observations exist that point to factors that are likely to be important but are obscure. They therefore deserve more research. I would like to mention some of them that strike me particularly.

I will use IQ as an example, just because there is some interesting material on straight environmental effects due to factors that potentially affect every genotype. The 5 IQ point handicap of twins is an example. It is well known that twins, irrespectively of whether they are fraternal or identical, are on average 5 IQ points lower than singletons; triplets are 9 points lower than singletons (McKeown and Record 1971). Counter to earlier belief that the source of this difference must be intrauterine competition between twins, observations by Record, McKeown and Edwards (1970) on twins whose cotwins died shortly after birth show that the effect is most likely to be postnatal. Is the decrease in parental attention per child responsible? Or is it the strong age-peer effect that twins reciprocally exercise one over the other and that tends to isolate them to some extent from the external world?

Probably the most important environmental effect detected comes from a study of adoptions (Skodak and Skeels 1949). One hundred children born to white mothers of average IQ 86 and adopted in "good" upper middle class families had an average IQ of 107. The IQ of the biological fathers is not known. Assuming random mating, and therefore an IQ of 100 for the father, the midparental IQ is $(86 + 100)/2 = 93$. The adopted children are then 14 IQ points above their biological parents. A small correction has been suggested to account for the fact that mothers are selected from a lower IQ group (see Jensen 1973); this is in part offset by possible nonassortative mating. It is difficult, however, to escape the conclusion that a change of environment, obviously an "improvement," can determine an effect of two-thirds to one standard deviation (one S.D. = 15 points of IQ).

What is it that makes an environment especially "good"? If it is a multitude of small factors, maybe it is not so interesting. But what if there is one single undetected environmental factor that is more interesting than the others? Observations on children conceived or born at the time of the Dutch famine (the winter of 1944-45) that show no effect on IQ eliminate short lasting nutritional deficits as an important environmental factor acting on IQ (Stein et al. 1972). As the authors recognize, this work does not affect conclusions on the effect of chronic severe malnutrition (total deficit or severe imbalance). One environmental effect, known earlier but now considerably clarified by elegant results on Dutch conscripts (Belmont and Marolla 1973), is the increasing handicap associated with increasing parity. This is a most interesting set of data that demands explanation on a psychological basis.

Other environmental influences may be important for developing certain special attitudes that are rare and desirable. They range from musical to other artistic attitudes, to scientific and to athletic ability. Society makes many efforts for identifying genotypes of promise, at least for athletic ability. Other types of skills are perhaps not so actively sought. At the lower end of the scale, some genotypes—a few of which, for example, Down's syndrome, are easily identified—may require special treatment in order to decrease the magnitude of the deficit that would usually result. But here identification of a specific condition and of a suitable treatment are usually difficult. The existence of wide, very probably heterogeneous syndromes (e.g., that of minimal brain damage) indicates the magnitude of the problems. That the differences involved are truly genetic can of course be stated only in a relatively small proportion of cases.

Whenever a genetic cause is detected, the source of the difference can in principle be reduced to some specific metabolic disturbance. Here may lie the key to treatment. Examples such as Tay-Sachs disease and the Lesch-Nyhan syndrome on one side are not susceptible to prophylaxis, other than by amniocentesis and abortion; but diseases like PKU (phenylketonuria), galactosemia, and others are, on the other hand, very encouraging examples of how biochemical genetics may help identification and treatment by environmental manipulation (McKusick and Claiborne 1973).

Having largely used IQ as an example, I should add that many psychologists reject the idea that our intellectual performance can be satisfactorily measured by a single quantity such as IQ, and my tendency is to agree with them that such an oversimplification obscures rather than enlightens. Again looking at the problem from a developmental point of view, I find approaches such as those by Piaget much more interesting. I believe that studies aimed at detecting genetic components by the Piagetian battery of tests should be encouraged. Naturally, such studies will suffer like any others from the inherent difficulties of analyzing genetic effects on behavior that I will mention later. Still, it would seem that the intellectual development of the child as studied by Piaget is a more interesting way of looking at behavior than just the measurement of a single quantity such as IQ or related measurements.

Geneticists are well aware that genes intervene at all stages of development and that much of development itself is essentially a complex program of turning on and off different genes. Eventually we may obtain a picture of the very complex program of development of an organism in terms of which genes are turned on and off, where and when. But this is still a difficult enough problem even for simple organisms and for the time being we have to be content with the general knowledge that gene regulation plays a very important part in our development. The best we can do now is to use situations such as the hemoglobin diseases for understanding the interplay of regulation and structural genes in other more complex processes. Such analogies are, however, rather remote and may not for the time being, at least, be very helpful.

Limitations of Present Genetic Research on Behavior in Man and Possibilities of Strengthening It

The classical genetic approach is a formal one. The first question a geneticist usually asks is whether one or more genes influence the particular trait that is being studied. In behavior even this very simple question is very often difficult to answer. Behavioral traits influenced by one gene (that are easier for the geneticist to study) are difficult to find. Moreover, there always is some significant environmental variation that tends to mask the single gene effect. Analysis of the inheritance of schizophrenia, a trait that should be especially amenable to investigation, shows that it is still very difficult to distinguish between a single gene hypothesis and a polygenic hypothesis in the presence of substantial environmental "background noise." In other cases such as IQ there is some evidence that a polygenic system is operating; even so, the establishment of the relative amount of background noise is difficult.

It has become customary to mask our ignorance and inability to proceed in this territory by making recourse to a quantity known as heritability. This is actually a family of quantities rather than a single quantity. Plant and animal breeders have used heritability to some advantage but the extension to analysis of human data is fraught with problems. There is frequent confusion between narrow and broad heritability. The two concepts have important differences. For breeders "narrow" heritability is of primary interest in that it allows one to predict the gain expected on artificial breeding. To the human investigator however, who does not plan artificial breeding for a better man, heritability has very limited utility. The tendency is then to use "broad" heritability, which is the proportion of the total variance that is due to genotypic differences. It is usually larger than the "narrow" estimate. But, especially in human populations, the assessment of the variance due to genotypic differences is fraught with great difficulties. First, the environment of different individuals is far from being randomized. Moreover, there is a variety of formulas, depending on which relatives are being measured and on the assumptions made, which tend to give different results. Finally, heritability estimates only the effect, at best, of these environmental differences that exist in the population sampled, which may be a trivial fraction of potential differences of practical importance. In the example of PKU, which has a heritability that is very close to 100 percent, phenotypes can be modified almost completely by environmental control. Thus, even a 100 percent heritability does not justify genetic fatalism; but people are willing to make sweeping statements and propose bold eugenic measures on the basis of an alleged heritability of 80 percent for IQ. The uncritical use of broad heritability tends to hide the fact that the environment can be more important than we think.

Schizophrenia provides another specific example that emphasizes the weaknesses of heritability measurements. If one considers schizophrenia as a polygenic disease, present estimates of heritability are of the order of 80 percent or higher. (The conditions under which heritability is measured in schizophrenia make it difficult even to state whether it is the narrow or broad heritability that is really being measured.) But if one uses the *same* data to compute heritability under two other models, both based on the hypothesis that there is a single major gene with environmental variation in liability, then one can obtain estimates of heritability of the order of 10 to 15 percent in one case, or up to 40 percent in the other case (Elston and Campbell 1970; James 1971; Kidd and Cavalli-Sforza 1973). In these two cases the assumptions and the computations are essentially the same. The difference lies in the scale used for measuring the trait. The first case uses a convention similar to that used for a polygenic trait, namely scores of 0, h, and 1 are assigned to the homozygote normal, heterozygote, and homozygote, with the higher liability for schizophrenia. The value of h is determined from the data. In the second case, one gives a score of zero to the normal phenotype and of 1 to the diseased. This is enough to change very substantially the estimate of heritability. One wonders whether it is not very misleading to base any conclusion on heritability measurements, at least when there is no clear indication of whether one or many genes are involved in the genetic determination.

If the first step is so difficult and we cannot even hope to measure the relative importance of environment and genetics in some convincing way, what further progress can one hope to make? It is important to note that the picture of possible advances in spite of this initial stumbling block is not as difficult as one might think. I believe in the power of two approaches. One of them is the study of linkage. We may be unable to say from a direct study of transmission in families whether a behavioral trait is polygenic or predominantly monogenic, but if we have additional information from linkage it may be possible to solve this question. There is already an example (manic-depressive psychosis) that shows evidence for sex-linked transmission, in part at least, confirmed by linkage with X-linked genes (Mendlewicz, Fleiss, and Fieve 1972; Reich, Clayton, and Winokur 1969). There is no reason why this kind of approach cannot be extended to autosomal genes. A truly polygenic trait, one in which there is a large number of genes, each with very small effect, will resist even this kind of analysis. However, there is always hope that some of the genes affecting complex polygenic traits will be more important than others, and so may be identified.

The limitation of the linkage approach at present is the relative poverty of genetic markers that have been mapped in man. A count of mapped genes as of February 4, 1974 indicated that some mapping information is available for at least 85 loci. Many of these markers, however, are not polymorphic because the linkage analysis has been based on interspecific differences in interspecific somatic cell hybrids, and therefore do not lend themselves easily to linkage analysis of a rare trait. A study by Jayakar (1970) has shown that with the information available a few years ago, the number of families to be studied for finding a linkage with known markers would be impossibly high, and even so there would be a relatively low probability of obtaining significant results. The picture is, however, changing rapidly, and more and more of the polymorphisms already known are pending a place on the genetic map. To date, six or seven chromosomes still do not have any markers unambiguously assigned to them. Thirteen groups of syntenic loci (whose linkage has been ascertained through family studies) have not yet been assigned to a chromosome. I would give a high priority to studies that increase our knowledge of polymorphisms and linkage, given that this is one of the ways in which the power of genetic analysis can be improved very substantially for the study of a genetically difficult organism such as man.

One even more basic way in which our genetic study of behavior can greatly profit is, I believe, from a deeper analysis at the biochemical level. A great number of research workers are concentrating at the moment on the enzymes and receptors in the brain. Unquestionably, their efforts will clarify considerably our understanding at least of abnormal behavior at the biochemical level. The study of psychotropic drugs has greatly reinforced the idea that certain metabolic pathways and certain substances in the brain are important for behavior. I find I learn more from reading that some individuals belonging to one family, who all showed schizophrenic symptoms, had a considerably reduced level of tetrahydrofolic reductase (Mudd and Freeman 1974) and benefited from the addition of folic acid to their diet than from any paper on the formal genetics of schizophrenia. The biochemical approach can be strengthened by extending the study from levels of enzyme activity to an analysis of qualitative differences between brain enzymes (Omenn and Motulsky 1972). There are many ways in which this can be done, electrophoretic techniques being one example. Immunological and physicochemical methods provide other ways to detect qualitative differences between enzymes or proteins that are almost certainly genetically determined. This approach may perhaps gain if carried at the level of enzymes involved in neurotransmitter metabolism (e.g., 5-hydroxy-tryptophan and dopa decarboxylase [Cavalli-Sforza et al. 1974]) or of other proteins having receptor or storage functions for such substances (Pignatti and Cavalli-Sforza 1974).

Genetic and Cultural Transmission in Behavioral Traits

Another approach seems to me relevant in the discussion of development and behavioral genetics, even if it opens up a whole series of new and superficially unrelated problems. I am referring to the study of cultural transmission, cultural inheritance, and cultural evolution. It seems to me almost self-evident that much behavior is transmitted culturally. By this I do not mean that genetic variation is negligible, but only that cultural transmission is a very important component in behavior. I also want to emphasize that this is probably the reason why there is such a discrepancy between the two schools of thought on behavior, namely the strictly environmental that has so far been dominant, and the genetic point of view that is creeping in, unfortunately not always in the most desirable way. It is remarkable that there is still so little common ground between these two extreme points of view. Hereditarians have often made the issue very difficult by the extraordinary claim that everything is so precisely inherited that there is almost no point in looking for environmental effects. Behaviorists on the other hand have made the opposite mistake. It is clear that the truth must be somewhere in between, but the problem is a little more subtle than just using heritability measurements to decide which of the two is more important.

I personally became interested in cultural transmission through recognizing that all types of evolution—cultural, biological, or otherwise—are driven by similar forces, which in biological evolution are called mutation, natural selection, and drift. In cultural evolution the equivalents of mutation are innovation and "copy error"; natural selection does not have a special name in cultural evolution, but those few anthropologists who have in recent times contributed to the discussion, not always very convincingly, have tended to use essentially the same words. The importance of drift in biological evolution has comparatively recently been the subject of renewed controversy as the recent discussion on molecular evolution has shown (Bodmer and Cavalli-Sforza 1971b). Drift is probably also important in cultural evolution, but the concept has found few supporters among anthropologists (Cavalli-Sforza 1971). Note that I am using the word "cultural" in a very general sense, encompassing a variety of factors not all, however, transmitted directly through chromosomes.

The major difference between cultural and biological evolution, apart from the most obvious one of the material that is evolving (genes for biology as compared with customs, ideas, and beliefs, in cultural evolution), is the nature of the transmission rules. We are quite confident about the rules of biological transmission. But the rules of cultural transmission are very poorly understood. This is surely a subject worthy of investigation.

Once the rules of transmission are specified, a mathematical theory becomes possible and may help in understanding cultural evolution, perhaps as much as it has helped in understanding biological evolution. I have made some preliminary efforts in this direction in collaboration with Mark Feldman at Stanford University. The model we have used is, perhaps not surprisingly, similar to an old model of biological transmission that has now been abandoned—the blending theory of inheritance suggested by Galton for explaining the correlations observed between relatives. In this theory both parents contribute an equal amount to their children and all the variation that arises is due to mutation. On recognizing that this would lead to a rather low variation, in contrast with the experience of all biologists, Fisher (1958) rejected the theory of blending inheritance and introduced the polygenic theory of inheritance. If we learned entirely from our parents, and to an equal extent from each with no underlying biological variation, blending inheritance would be the natural theory of cultural transmission. It so happens that cultural transmission is much more complicated than this and not only parents but many other people contribute to our cultural development. Starting with these very simple assumptions, we have shown how it is possible to predict the changes due to mutation and effects of finite population size, that is drift, in populations subject to cultural transmission mediated by the effect of parents alone, and also by the effects of teachers or "leaders" (Cavalli-Sforza and Feldman 1973a; Feldman and Cavalli-Sforza 1974). We have also shown that there are differences in evolutionary rates and in individual variation when the transmission is mostly through age peers, or parents alone, and that earlier generations also have an effect.

Our analysis has only begun, but I would like to summarize some conclusions that seem reasonable at this stage. One important conclusion (that could have been reached by the simple recognition that the transmission of culture, and with it behavior, is of the blending type) is that the variation between individuals of the same social group for a culturally transmitted trait will be, under similar circumstances, lower than if the trait is transmitted biologically. This fits nicely with the necessity for low variation in a number of cultural and behavioral traits. Take language as an extreme example. We could not expect to understand each other without considerable homogeneity in the language spoken by people within a given population. This homogeneity is made possible by a considerable plasticity of individuals and a large amount of feedback from parents, teachers and age peers in the learning process. Beliefs and social customs or traits that one may be willing to call behavioral (e.g., sex behavior) follow a similar pattern of transmission.

Another important consideration is the overwhelming effect of certain individuals, be they teachers or social or political leaders, in determining people's behavior. The mere fact that there can exist teachers and social leaders increases homogeneity within the group, but it also increases the variation between groups led by different leaders. It has a similar effect on the same group over time, when the influence of one leader is overruled by that of another. Thus, one can expect for certain cultural (including some behavioral) traits considerable homogeneity over a given area and period, but also the possibility of radical changes from time to time. All this has been put into admittedly crude formulas. It would be important to show applications of these formulas to real data and this has not yet been done.

In the simple theory of cultural inheritance that I have summarized briefly, we have initially considered all individuals to be identical genetically. A given trait in an individual is expressed as the weighted average of the traits of the people who have contributed to his formation, where the weights indicate the relative contribution of each individual. In addition, there is a random "copy error" that might include innovation. The initial model therefore assumed an equal capacity of all individuals to absorb cultural information (that is, equal "plasticity") and equal exposure to external influences (constituting the "environment" in which development takes place—see Cavalli-Sforza [1974]). This rather severe restriction can be removed. In particular, as geneticists we are interested in taking into account inherited differences in individual plasticity. The genotype determines the potential response to the environment and the phenotype is the realization of this potential. These usual definitions, found in most textbooks of genetics, had not previously been formalized mathematically so as to take into account both cultural and biological transmission. There are mathematical difficulties; nevertheless, an attempt has been made to develop a model with just one two-allele locus (Cavalli-Sforza and Feldman 1973b) and with cultural transmission only from parents to children. "Cultural" transmission may be misunderstood, given that there is no wide agreement on definitions of culture. In the model, nongenetic transmission was called "phenotypic" and was designed to cover all types of transmission that do not take place directly through the chromosomal mechanism: It includes therefore all environmental effects that are transmissible over generations, ranging from those that might be called more strictly cultural, to others that might be more related to the physical environment. Even many factors of the physical environment, however, may have a cultural component (e.g., diet). This approach is being extended and one may then think about the problem of estimation from real data, which has not been seriously considered so far.

One of the conclusions from this research has already been stated a number of times. It is that there is no way to distinguish between cultural and biological transmission unless one can study adoptions and test the similarity with *both* biological *and* adoptive relatives. A table calculated for the expected correlations between relatives, adopted and nonadopted (Cavalli-Sforza and Feldman 1973b) restates this well known principle in a dramatic form, and shows some generalizations that are possible. In the absence of adoption studies there is no hope of distinguishing rigorously whether standard measurements of inheritance, that is similarities between relatives (of any kind), are due to genetic determination of the trait differences, or to sociocultural inheritance (more generally, phenotypic transmission), or to a mixture of the two because correlations between relatives are similar in both models. The regression of adopted children on adoptive parents gives a direct estimate of the children's plasticity with respect to phenotypic transmission from the parents. The regression or correlation of adopted children with biological parents is not as easy to interpret as is usually believed, as shown by the complexity of the terms that appear in them.

It is very unfortunate for the geneticist that adoptions are relatively rare events. Also the criteria that are used by adoption agencies often give rise to biases such as the well known tendency for selective placement exercised by many agencies. Their effects on the correlations are difficult to determine, but are sufficient to throw some doubt on the estimates obtained. But I have a feeling that there are important untapped resources in this area. There are social customs in certain populations which make adoption a common event. In particular I want to refer to a Chinese custom still practiced until quite recently in Taiwan, although it has almost disappeared on the mainland. One of these customs involves people from families that are not very wealthy who have their daughters adopted by other families shortly after birth, so that they will be able to marry the male children of the adopting families when the girls are grown up. This involves a financial reward for the family that gives up the child and a financial advantage for the family that receives the child (bride prices for adult girls being higher). Some psychological consequences of this custom are being studied by Wolf (1975). The custom is now disappearing, but there still are substantial numbers of people of age eighteen years or even less who have been raised in such adoptive families. Biological parents and sibs are known and are available for study, as well as the adoptive parents and sibs. In most other adoption studies biological parents are not always available, which severely limits the opportunities for analysis. There is some bias here due to the fact that adopted girls tend to carry a social stigma. Controls, in terms of biological and adoptive sibs are, however, available for study.

I am sure that the world of anthropology can offer other important sources of adoption material that could be useful for distinguishing the effects of cultural and biological inheritance. Naturally, they may give information on sociocultural environments that differ too much from our own for a direct carry over of the conclusions obtained, but I do not believe this should deter us from studying them. Also in the Western world there are related sources, such as children reared in nurseries, which have been used to a very limited extent. A recent preliminary report by Tizard (1974) compares the IQ of black, racially mixed, and white children brought up in good nurseries. The only significant difference found is in favor of black and racially mixed children. These studies, if confirmed and extended, can rule out Jensen's hypothesis that the black-white difference in IQ is genetic.

The last subject that I will briefly touch on is the problem of what one would like to know from a practical point of view. Among the interesting components of the variance of a trait are genotype-environment interaction and covariance. Briefly, the first means that different genotypes react differently in different environments and the second that different genotypes are found in different environments. Actually, it is not clear what fraction of such genotype-environment interactions and covariances enter into the usual estimates of broad heritability and which are left out. Even if these interactions and covariances could be estimated easily, there would remain a desire to dissect these quantities in more useful ways than just estimating their total amount.

What we would really like to know is how given genotypes react in given environments. This seems in itself an almost hopeless task, given the enormous variation in genotypes, the great possible variation of environments and difficulty of doing any carefully controlled tests on the effects of the environment on a given genotype. Even if this looks like an impossible dream, some possibilities are open. It is here that twin research may have a considerable impact, much greater than that of estimating heritabilities. Twins are different from the rest of the population in a number of ways and many of the conclusions that are obtained from them are difficult to extrapolate to the rest of the population. Among the possibilities: Knowledge might be gained by having monozygous twins follow, for instance, different types of scholastic curricula. This kind of information is of course very difficult to obtain. The problems that one faces are enormous. For one thing didactic techniques are not sufficiently well standardized and differentiated (with few exceptions), and this kind of experiment may be very difficult.

The typology of behavior of personality is sufficiently underdeveloped that it may be very difficult to extrapolate conclusions from some pairs of twins to the rest of the population. Maybe such a typology is even inconceivable, given the enormous number of different genotypes. Still a few genes—or at least phenotypic traits—may be more important than others, making some generalization possible. Still more important, this research involves a moral problem in experimenting on human subjects: Behavioral investigations are similar to clinical ones in that one has to be very careful that the research done in no way damages the subject on which the research is being carried out. One consideration here is that some psychologists suggest that twins may benefit if differences between them rather than their similarities are emphasized.

In summary, the task before the human behavior geneticist is a difficult one. It is, however, not impossible and certain avenues seem more profitable than others. We will have to get rid of some of our present fads and fallacies, like the idea that establishing a measurement of heritability is all that is necessary, or to ignore the existence of cultural transmission, as has been the custom with most students of human behavioral genetics, or similarly to ignore the existence of genetic variation as has been the unfortunate practice of the behaviorists.

References

Belmont, L., and Marolla, F.A. 1973. Birth order, family size and intelligence, *Science* 182: 1096-1101.

Bodmer, W., and Cavalli-Sforza, L.L. 1970. Intelligence and race. *Scientific American* 223 (October): 19-29.

———. 1971a. Fear of enlightenment. *Nature* 229: 71-72.

———. 1971b. Variation in fitness and molecular evolution. *Proceedings of the Sixth Berkely Symposium on mathematical statistics and probability*, pp. 155-175. Berkeley: University of California Press.

Cavalli-Sforza, L.L. 1971. Similarities and dissimilarities of socio-cultural and biological evolution. In *Mathematics in the archaeological and historical Sciences,* eds. F. R. Hodson, D.G. Kendall, and P. Tautu, pp. 535-41. Edinburgh: Edinburgh University Press.

_____. 1974. The role of plasticity in biological and cultural evolution. *Annals of the New York Academy of Science* 231: 43-59.

Cavalli-Sforza, L.L., and Feldman, M.W. 1973a. Models for cultural inheritance I. Group mean and within group variation. *Theoretical Population Biology* 4: 42-55.

_____. 1973b. Cultural versus biological inheritance: Phenotypic transmission from parent to children. *American Journal of Human Genetics* 25: 618-37.

Cavalli-Sforza, L.L., Santachiara, S.A., Wang, L., Erdelyi, E., and Barchas, J. 1974. Electrophoretic study of 5-hydroxytryptophan decarboxylase from brain and liver in several species. *Journal of Neurochemistry* 23: 629-34.

Elston, R.C., and Campbell, M.A. 1970. Schizophrenia: Evidence for the major gene hypothesis. *Behavior Genetics* 1: 3-10.

Feldman, M., and Cavalli-Sforza, L.L. 1974. Models for cultural inheritance: A general linear model. Unpublished paper, Stanford University.

Fisher, R.A. 1958. *Genetical theory of natural selection*, 2nd ed., New York: Dover Publications.

James, J.W. 1971. Frequency in relatives for an all-or-none trait. *Annals of Human Genetics* 35: 47.

Jayakar, S.D. 1970. On the detection and estimation of linkage between a locus influencing a quantitative character and a marker locus. *Biometrics* 26: 451-64.

Jensen, A.R. 1973. Let's understand Skodak and Skeels, finally. *Educational Psychologist* 10: 30-35.

Kidd, K.K., and Cavalli-Sforza, L.L. 1973. An analysis of the genetics of schizophrenia. *Social Biology* 20: 254-65.

McCall, R.B. 1970. Intelligence quotient pattern over age: Comparisons among siblings and parent-child pairs. *Science* 170: 644-48.

_____. 1972. Similarity in developmental profile among related pairs of human infants. *Science* 178: 1004-5.

McKeown, T., and Record, R.G. 1971. Early environment influences on the development of intelligence. *British Medical Bulletin* 27: 48-52.

McKusick, V.A., and Claiborne, R. 1973. *Medical genetics*. New York: HP Publishing Co.

Mendlewicz, J., Fleiss, M., and Fieve, R.R. 1972. Evidence for X linkage in transmission of manic depressive illness. *Journal of the American Medical Association* 222: 1624.

Mudd, S.H., and Freeman, J.M. 1974. $N^{5,10}$-methylenetetrahydrofolate reductase deficiency and schizophrenia: A working hypothesis. In press.

Nature. 1970. Fear of Enlightenment. *Nature* 228: 1013-14.

Omenn, G.S., and Motulsky, A.G. 1972. Biochemical genetics and the evolution of human behavior. In *Genetics, environment and behavior: Implications for educational policy,* eds. L. Ehrman, G.S. Omenn and E. Caspari. New York: Academic Press.

Pignatti, P.F., and Cavalli-Sforza, L.L. 1974. Serotonin binding proteins from human blood platelets. *Neurobiology*. In press.

Record, R.G., McKeown, T., and Edwards, J.H. 1970. An investigation of the difference in measured intelligence between twins and single births. *Annals of Human Genetics, London* 34: 11-20.

Reich, T., Clayton, P.J., and Winokur, G. 1969. Family history studies: V. The genetics of mania. *American Journal of Psychiatry* 125: 1358-69.

Sears, R.R., and Feldman, S.S. 1964. *The seven ages of Man*. Palo Alto, Calif.: William Kauffman, Inc.

Skodak, M., and Skeels, H.M. 1949. A final follow up study of one hundred adopted children. *Journal of Genetic Psychology* 75: 85-125.

Stein, Z., Susser, M., Saenger, G., and Marolla, F. 1972. Nutrition and mental performance. *Science* 178: 708-13.

Tizard, B. 1974. IQ and race. *Nature* 247: 316.

Wilson, R.S. 1972. Similarity in developmental profile among related pairs of human infants. *Science* 178: 1006-7.

Wolf, A. 1975. The women of Hai-Shan: A demographic portrait. In *Women in Chinese society,* eds. M. Wolf and R. Witke. Palo Alto, Calif.: Stanford University Press. In press.

Commentary I

*I. Michael Lerner
University of California
at Berkeley*

One of the purposes of the workshop that led to this book was to aid the National Institutes of Health (NIH) study section particularly concerned with developmental human behavior genetics, to identify the promising directions in this field, and to formulate the criteria by which proposals for studies in this area may be evaluated. Cavalli-Sforza's discussion in this chapter provides much food for thought along these lines, but has also left some latitude for disagreement. From a discussant's point of view this is a very fortunate circumstance. One need not be a Hegelian or a Marxist to believe that out of a clash of opposite opinions a fruitful synthesis may arise. It is precisely in such dialectical spirit that I would like to comment on some of the points Cavalli-Sforza made. It would, of course, be foolhardy to attempt to fault him on substantive matters; hence, these remarks are addressed largely to some reasonably broad issues.

He has identified five general areas of techniques of behavior genetics that appear to be the most feasible and potentially rewarding to pursue:

1. Longitudinal inquiries on human behavior
2. Studies discriminating between monogenic and polygenic inheritance of behavior traits by using linkage and mapping methods
3. Deeper analysis of behavior on the biochemical level
4. Investigation of cultural transmission
5. Study of genotype-environment interaction

With respect to the first, that is, the need for longitudinal and projective studies, one can only concur wholeheartedly. Investigation of development after the fact is not very productive. There is little doubt that longitudinal study, whether biochemical or psychological, is an absolute essential if we are ever to approach an understanding of human behavior in its genetic and other aspects.

But disagreement with one of Cavalli-Sforza's assumptions regarding the fundamental strategy of longitudinal investigations is possible. Being in the Sixth Age of Man, he is bothered by the alleged brevity of his remaining research career, brevity that he feels would prevent him from making a significant contribution along these lines; hence, his recommendation is that it is the younger people who should be urged to invest their efforts in long-term investigations needed to pursue this approach. I, on the contrary, feel that it is us graybeards who should initiate such projects. Being well into the Seventh Age, I am not in the least concerned about anything accruing to me personally from the results of whatever study I may begin. This is not mere lip service to youth. As any ex-animal breeder knows, one of the main reasons that only a few really bright young biologists enter the field of large animal genetics is the unconscionable long time before a research project on such material can lead to publication of other than trivial results. It is doubtful that all the emeralds in the King of Thailand's treasury would tempt promising young geneticists to devote themselves to the study of color inheritance in Siamese elephants, unless, that is, someone also had already laid the ground for the ultimate payoff in research production. Hence, it is incumbent on the somewhat older workers to prepare the material for the usufruct of their successors, whenever long periods of data gathering have to precede interpretations. Indeed, this was exactly the spirit in which a psychologist colleague of roughly my own age and I designed a longitudinal study of certain aspects of twin behavior. Our role was to start the investigation and then consign it to more youthful workers, who could, so to say, come in on the kill, perhaps, even after we ourselves passed beyond the Seventh Age of Man. It may be added that the project, having gone for study to a less enlightened section of a different institute than this one, was not funded. But all the same, long-range studies might well be designed by experienced hands whether or not they participate in the final analysis.

The other point that might be noted in reference to Cavalli-Sforza's discussion of this area is that the study of Dutch conscripts to which he refers is probably not the last word on the effect of maternal nutrition on the IQ of children. Why this may be so is discussed in some detail in Shneour's (1974) recent book *The Malnourished Mind*, which has considerable documentation on the matter.

The second area of Cavalli-Sforza's concern is the investigation of the formal genetics of behavior traits. Here one can voice complete agreement with his emphasis on the need to clarify the persistent confusion in the minds of many students of behavior as to the meaning of the family of quantities referred to as heritability. The term originated in Lush's laboratory at a date that, in spite of the fact that the people involved are still with us, seems to be hard to pin down. In any case the concept underwent various refinements and transformations but still appears often to be applied in unwarranted ways. Cavalli-Sforza's point is that estimates made even by sophisticated and fully informed quantitative geneticists may suffer from ambiguity depending on underlying models. When environment is not randomized (as it never is in humans) estimates can vary to the degree that they become meaningless, as has been demonstrated in the little appreciated but, to my mind, classical paper on cattle twins by Donald (1959) that he presented over fifteen years ago at the Montreal International

Genetics Congress, and that, incidentally, figures in Layzer's (1974) recent analysis of the informational content of heritability.

But aside from the precise evaluation of heritability in narrow and broad senses, there also seems to be a great deal of confusion among behavior scientists about its general meaning. Thus, in a recent book, Eysenck (1971) writing about IQ explains that: "The figure of 80 percent heritability is an average. It does not apply equally to every person in the country. For some people environment may play a much bigger part than is suggested by this figure; for others it may be even less." The subversion of a population concept to "apply" to single individuals is certainly an example of ruin upon ruin, rout on rout, confusion worse confounded.

The complications of determining heritability of such traits as schizophrenia have been clearly illustrated by Cavalli-Sforza. It might also be emphasized that in general the study of formal genetics of such characters is beset by the lack of a proper taxonomy. To recall once more the situation in animal breeding, the power of the methods of studying polygenic inheritance was demonstrated there only after the traits to be investigated or selected for were clearly identified. A good many of the monogenically determined behavior pathologies of man have attained this status, but I do not feel that schizophrenia has. I therefore suspect that diagnostic criteria more generally agreed upon and more subject to validation than they seem to be at present are needed before real progress on the formal genetic front will be possible. Furthermore, welcome as therapeutic advances are in this area, they do complicate investigations of the hereditary basis of schizophrenia, a fact which on the balance is not a loss.

Cavalli-Sforza's third main point, that is, the need for study of behavior genetics in depth, is also a hardly debatable issue. But a word of caution is necessary here. Doing biochemical or molecular research is no guarantee per se of productive results, as is properly implied in his reference to biochemical studies of behavior, which ignore the genetic approach. One of the perils to be guarded against lies in the enthusiastic but indiscriminate proliferation of ultimate, more ultimate, and most ultimate solutions of the riddles in the biochemistry of behavior pathologies. For schizophrenia there seems to be a procession of diagnostic or causal substances identified in blood serum or urine. They have succeeded each other in an endless sequence probably for over a quarter of a century. The thesis that schizophrenia is a physiological deficiency identifiable from metabolic patterns goes back at least to studies of biochemical individuality in Roger Williams' laboratory (Young et al. 1951). I do not know how one is to judge the promise of one or another lead in this direction. Somehow one has to navigate within plausible channels, avoiding bizarre extremes, but it would take a great deal more biochemical competence than I have even to suggest what reasonable evaluating procedures may be.

We now come to a more controversial point of Cavalli-Sforza's conspectus. He is currently engaged in a very important formulation involving cultural evolution. The models that he is developing are fascinating and potentially fruitful for the understanding of man's past and present and the guiding of his future. Yet, I confess to feeling somewhat uneasy about this subject, at least in the present context. First, reasoning by analogy from one field to another, from biology to anthropology and sociology in this particular case, is always a risky business. The fact that the contrast between Mendelian inheritance as against blending transmission is being considered alleviates the problem of comparing evolutionary mechanisms at different levels of organization only in part. Not only is there danger of oversimplification but there are also possibilities of real misrepresentation that can lead to unhappy consequences.

The inglorious chapter of our cultural history called Social Darwinism has been told often enough, among others by Hofstadter (1955). What starts out as a *descriptive* theory can admirably become a *prospective* theory, but it can also wind up as a *prescriptive* theory. Various generalizations about cultural evolution involving transplantation of such mechanisms of organic evolution as mutation, migration, and selection (even if it does not appear in Cavalli-Sforza's scheme) into cultural history have been made probably by many, at least since Sir Edward Tylor (who incidentally vigorously denied that his notions derived either from Darwin or from Herbert Spencer). The great advance being made by Cavalli-Sforza and Feldman is their attempt at rigorous model construction, quantification and projected validation.

Yet, much as the intellectually alluring area of the interaction between organic and cultural evolution intrigues me, there seems to be little connection between it and developmental behavior genetics. One may argue that an all-embracing view of this specific field of research might include the clarification of the bases of *cultural* transmission (it goes without saying that social transmission is part of individual developmental history). True enough, the significance of such enterprises should not be underestimated. But, nevertheless, sidetracking the resources available for the study of development in a strict sense to studies in cultural anthropology, provocative and consequential as they may be, might be a disservice to the goals of our present deliberations.

There is also one rather trifling point in reference to this section of Cavalli-Sforza's discussion that should be made if only for the sake of the record. One of his statements reads that it was R. A. Fisher who "rejected the theory of blending inheritance and introduced the polygenic theory of inheritance." Surely Fisher (and this was many years before he cast doubts on the integrity of Mendel's data) rejected blending inheritance in 1918 on the basis of Mendel's experiments with peas. And just as surely, Mendel in reporting his experiments with beans and his observations on color in ornamental plants anticipated Fisher by half a century in advancing the notion of polygenic inheritance.

Cavalli-Sforza's final recommendation deals with the study of genotype-environment interaction. This is an area in which as much misunderstanding exists as in considerations of heritability. For example to cite Eysenck (1971) once more, he obviously also confounds genotype-environment correlation with genotype-environment interaction, when he says that "our calculations disregard the importance of interaction between environment and heredity. Thus, intelligent children may select a different environment to grow up in than do dull children; this different environment in turn increases the difference in I.Q."

The genotype-environment *correlation* reflects the changes in the environmental component of the phenotype that accompany the changes in the genetic component (or vice versa). *Interaction,* however, refers to the differential effects of specific genotypes in different environments, such as explored by Haldane (1946) and has nothing to do with genotype-dependent selection of environment. We should, of course, know about the role and magnitude of these quantities in determining human behavior. This topic most assuredly deserves the emphasis given to it in the paper under discussion.

Finally, I might comment on a general issue with which Cavalli-Sforza introduced his presentation, that is, the ethics of investigation of human behavior. To state the obvious: We have to steer a tricky course between the values of gathering knowledge and the hazards, both of violating human rights and of arriving at unpalatable conclusions. Presumably pursuit of truth unavoidably carries with it the latter risk, and we have to live with it. Somewhat less exciting is the prospect of facing the unpredictable consequences of emotionally motivated half-baked and irresponsible pronouncements on whatever side of a disputable question they are made. I might, however, conclude with just a single example bearing on this problem that should illustrate the unavoidability of subjective priorities, if it does nothing else.

I have no idea what the precise heritability value for human intelligence is. Even if I did, I would not know what to do with it, beyond continually striving to disabuse the laity of the common notion that high heritability and ease of environmental modification are mutually exclusive. I have even less knowledge of and interest in the very likely well-meaning, but to me naive, attempts to decide on the precise proportion contributed by genetic differences in this trait to phenotypic variance. What I do know is that (a) I am becoming exceedingly weary of endless regurgitation of topics that cannot be adjudicated by currently available data or techniques without some preposterous assumptions; (b) there is no such thing as too much knowledge, and competent research efforts in any area, whether it bores me personally or not should be encouraged; and finally, (c) there are any number of research projects in the areas outlined by Cavalli-Sforza that are worthy of support without raising the hackles of partisans of one or another hypothesis that may have little or no bearing on the harvest of biological information expected from such projects.

References

Donald, H.P. 1959. Evidence from twins on variation in growth and production of cattle. *Proceedings of the 10th International Congress of Genetics, Montreal* 1: 225-35.

Eysenck, H.J. 1971. *The I.Q. argument.* New York: Library Press.

Haldane, J.B.S. 1964. The interaction of nature and nurture. *Annals of Eugenics* 13: 197-205.

Hofstadter, R. 1955. *Social Darwinism in American thought.* Rev. ed. Boston: Beacon Press.

Layzer, D. 1974. Heritability analyses of IQ scores: Science or numerology? *Science* 183: 1259-66.

Shneour, E.A. 1974. *The malnourished mind.* Garden City, N.Y.: Anchor Press/Doubleday.

Young, M.K. Jr.; Berry, H.K.; Beerstecher, E., Jr.; and Berry, J.S. 1951. Metabolic patterns of schizophrenic and control groups. *Biochemical Institute Studies* 4: 189-97. University of Texas Publication No. 5109.

Commentary II

John C. DeFries
University of Colorado

In his excellent, far-ranging discussion, Professor Cavalli-Sforza considers many important theoretical and methodological issues within the field of human behavioral genetics. Since a scholarly critique of his comments has already been provided by Professor Lerner, however, only one of the numerous significant points raised by Professor Cavalli-Sforza will be enlarged upon in the present discussion—the essential role of adoption studies in human behavioral genetics research.

While recently reviewing the literature of human behavioral genetics with G. E. McClearn (McClearn and DeFries 1973), I was particularly struck by the power of the adoption method. In order to illustrate this power three relatively recent adoption studies are reviewed briefly. The first is the classic adoption study of schizophrenia by Heston (1966): Heston's experimental group consisted of young adults born to schizophrenic mothers, but permanently separated from them during the first month of life and reared in foster or adoptive homes. For controls, adoptees or foster children born to mothers with no record of psychiatric disturbance were utilized. Heston assessed the behavior of these experimental and control subjects by various means—psychiatric interview, and reviews of school, police, medical, and Veterans Administration records. The results of this study are summarized in Table 6-1. Although there are a number of indications of increased psychopathology among the experimental group, note especially that there were five diagnosed cases of schizophrenia among the experimental subjects versus none among the controls. More recently, similar evidence has been found by Kety and Rosenthal (Kety et al. 1971; Rosenthal et al. 1971), using variations of the adopted-child method in Denmark.

Using a similar approach, Crowe (1972) studied the antisocial behavior of adopted children whose biological mothers were convicted criminals incarcerated in the state of Iowa. A control group consisted of adoptees matched to the probands for age, sex, race, and age at the time of adoptive decree. Crowe obtained data concerning the incidence of antisocial behavior of subjects from records of the Iowa Bureau of Criminal Investigation. This state agency maintains records of every adult arrest in which the individual is officially charged, as well as records of juvenile offenders sent to state training schools. The results of this study are summarized in Table 6-2. Note that eight probands accounted for a total of eighteen arrests, versus only two among controls. In addition, five probands were incarcer-

Table 6-1
Results of a Study of Persons Born to Schizophrenic Mothers and Reared in Adoptive or Foster Homes, and of Controls Born to Normal Parents and Similarly Reared

Item	Control	Experimental	Exact Probability (Fisher's Test)
Number of subjects	50	47	
Number of males	33	30	
Age, mean (years)	36.3	35.8	
Number adopted	19	22	
MHSRS, means[a]	80.1	65.2	.0006
Number with schizophrenia	0	5	.024
Number with mental deficiency (IQ < 70)	0	4	.052
Number with antisocial personalities	2	9	.017
Number with neurotic personality disorder	7	13	.052
Number spending more than one year in penal or psychiatric institution	2	11	.006
Total years incarcerated	15	112	
Number of felons	2	7	.054
Number serving in armed forces	17	21	
Number discharged from armed forces on psychiatric or behavioral grounds	1	8	.021
Social group, first home, mean	4.2	4.5	
Social group, present, mean	4.7	5.4	
IQ, mean	103.7	94.0	
Years in school, mean	12.4	11.6	
Number of children, total	84	71	
Number of divorces, total	7	6	
Number never married, > 30 years of age	4	9	

[a]The MHSRS is a global rating of psychopathology moving from 0 to 100 with decreasing psychopathology. Total group mean, 72.8; S.D., 18.4.

Source: L.L. Heston (1970, pp. 249-256).

ated for a total of three and one-half years, whereas none of the controls was ever incarcerated.

The third study, conducted in Denmark, concerns the incidence of drinking problems among adoptees with an alcoholic biological parent. As probands, Goodwin and colleagues (1973) identified a group of fifty-five adult males who had had a biological parent with a hospital diagnosis of alcoholism, but who had been permanently separated from these biological parents prior to six weeks of age and placed with nonrelatives. A control group consisted of seventy-eight similarly adopted males with no known alcoholism in their biological parentage. A strict criterion of alcoholism was used in this study: in addition to heavy drinking, evidence of social or marital problems, difficulties with employment or the law, and various physiological symptoms were required. Results of this study are summarized in Table 6-3. Although it may be seen that both probands and

Table 6-2
Arrest Records of the Biological Offspring (Reared by Adoptive Parents) of Female Criminal Offenders Compared with Those of Controls

Arrest Records	Probands	Controls	p[a]
Number of subjects checked for records	52	52	
Subjects with records	8	2	.046
Total number of arrests	18	2	
Subjects arrested as adults	7	2	.084
Subjects with convictions	7	1	.030
Subjects with two or more arrests	4	0	.059
Subjects incarcerated for an offense[b]	5	0	.028
Total time incarcerated, years[b]	3.5	0	

[a]Fisher's exact test.
[b]This includes two subjects sent to training school as juveniles, which accounts for 1.5 years.
Source: R.R. Crowe (1972, pp. 600-603).

controls were rather heavy drinkers in this population, 18 percent of the probands were diagnosed as alcoholic, versus only 5 percent among controls.

The three studies reviewed above each provides convincing evidence for the presence of a heritable component in the characters under study. However, each of these studies has been *retrospective*, and thus suffers certain methodological inadequacies (incomplete information on biological and adoptive parents, etc.). In spite of these inadequacies, retrospective adoption studies have an important place in human behavioral genetics and clearly are worthy of continued support. Nevertheless, the time has come to also consider the merits of a *prospective* adoption study. In a prospective study, adoptive mothers, adoptive fathers, biological mothers, and even in some cases biological fathers could be administered a comprehensive battery of behavioral tests. Subsequently, data could be obtained from normal adoptees, as well as from those "at risk" for various characters of interest. Although longitudinal studies are rather unpopular with funding agencies, it seems to me that if a prospective adoption study is ever undertaken, it would be essential for at least some of the children to be tested at several different ages to assess for possible developmental differences.

Data from a prospective adoption study could provide unambiguous evidence for the presence (or absence) of a heritable component in each of the behaviors under study. In addition, if comparable data were obtained on children reared by their biological parents, the relative roles of cultural versus biological inheritance could be assessed in the manner outlined by Professor Cavalli-Sforza. Such an undertaking would clearly represent a

Table 6-3
Comparison of Drinking Problems and Patterns in Two Adoptive Groups (%)

Item	Probands (N = 55)	Controls (N = 78)
Hallucinations[a]	6	0
Lost control[b]	35	17
Amnesia	53	41
Tremor	24	22
Morning drinking[a]	29	11
Delirium tremens	6	1
Rum fits	2	0
Social disapproval	6	8
Marital trouble	18	9
Job trouble	7	3
Drunken driving arrests	7	4
Police trouble, other	15	8
Treated for drinking, any[a]	9	1
Hospitalized for drinking	7	0
Drinking pattern		
Moderate drinker	51	45
Heavy drinker, ever	22	36
Problem drinker, ever	9	14
Alcoholic, ever[b]	18	5

[a] $P < .05$.
[b] $P < .02$.
Source: D.W. Goodwin et al. (1973, pp. 238-43).

long-term commitment. Nevertheless, due to the potentially great value of such a study to the field of human behavioral genetics, it is at least worthy of consideration.

Immediate attention should be given to the merits of a prospective adoption study. Due to the increasingly widespread use of contraceptives among young people, as well as access to legalized abortion, we may be witnessing the last generation of adoptive placement of young illegitimate children in this country. Consider some data recently provided by the Colorado State Department of Social Services: During the past four years the Colorado County Departments of Social Services participated in the placement of the following numbers of children under one year of age: In 1970, 888 children were placed; in 1971, 581; in 1972, 406; and in 1973, 312. (These data were obtained from Program Evaluation Report, CS 73-1, and from A. Snook, personal communication.) According to the adoption consultant for this agency, this pattern is not unique to Colorado—it is being observed throughout the country. Thus, if a prospective adoption study is ever to be undertaken with a United States population, it may be necessary to do so within the relatively near future.

References

Crowe, R. R. 1972. The adopted offspring of women criminal offenders: A study of their arrest records. *Archives of General Psychiatry* 27: 600-603.

Goodwin, D. W.; Schulsinger, F.; Hermanson, L.: Guze, S. B.; and Winokur, G. 1973. Alcohol problems in adoptees raised apart from alcoholic biological parents. *Archives of General Psychiatry* 28: 238-43.

Heston, L. L. 1966. Psychiatric disorders in foster home reared children of schizophrenic mothers. *British Journal of Psychiatry* 112: 819-25.

Heston, L. L. 1970. The genetics of schizophrenic and schizoid disease. *Science* 167: 249-56.

Kety, S. S.; Rosenthal, D.; Wender, P. H.; and Schulsinger, F. 1971. Mental illness in the biological and adoptive families of adopted schizophrenics. *American Journal of Psychiatry* 128:302-6.

McClearn, G.E., and DeFries, J.C. 1973. *Introduction to behavioral genetics*. San Francisco: W.H. Freeman and Co.

Rosenthal, D.; Wender, P.H.; Kety, S.S.; Welner, J.; and Schulsinger, F. 1971. The adopted-away offspring of schizophrenics. *American Journal of Psychiatry* 128: 307-11.

7

Counseling in Genetics and Applied Behavior Genetics

John Money
The Johns Hopkins
University

Genetic counseling as it is known today is really three entities: education, counseling, and psychotherapy as related to genetic transmission. Among clinical geneticists there has been an erroneous tendency to equate genetic counseling with genetic education, as though on the assumption that the carriers of bad genes or defective chromosomes will be able to obey the intellectual and logical dictates of statistical genetics. But copulating and conceiving are subject to far more complexity of personal decision than can be accounted for by statistical logic. There may, indeed, be some sturdy couples who are able to assimilate the geneticist's statistics and apply them to their own personal lives and decisions, if not alone, then with the help of friends and relatives. Others will need professional genetic counseling to help them resolve the shock of the personal implications of the genetic education imparted to them before they can assimilate and apply it. Still others, psychologically vulnerable, will be in so severe a state of shock, either at having produced a genetically defective child, or of having learned of their prospect of doing so, that they will need not simply personalized counseling but a more extensive program of psychotherapy.

Psychotherapy will require priority in some instances before educational information can be assimilated. In some cases the need for psychotherapy will arise from a person's own state of psychiatric vulnerability, or in response, perhaps exaggerated, to knowledge of his or her genetic dilemma. In other cases, the need may arise iatrogenically, for there is no doubt that genetic education given by a person not specifically trained as an expert in genetic counseling might do more harm than good. For example, prematurely identifying a husband or wife as the carrier of a dominant gene may destroy their relationship if it is already neurotically unstable, and possibly precipitate a suicidal depression. In couples not already showing signs of genetics shock, the words and concepts of genetic education are important in preventing such shock.

Preparation of this chapter was supported in part by USPHS grant HD00325 and by funds from the Grant Foundation.

Genetic Education

For a lesson in personal genetics to be assimilable, the teacher needs first of all to ascertain what conception of heredity his student may already have in mind. It goes without saying that one's formulations for a graduate in biology and for a grade school dropout will be differently worded.

For the averagely informed person untutored in scientific genetics, heredity means family-tree heredity. That is to say, one inherits a trait or condition from a forebear in much the same way as one inherits property or a debt without having personal control over the conditions of the will. People vary in their beliefs as to whether bad heredity can be circumvented. Some have a gambler's philosophy that good luck will always smile on them. Some believe that God can intervene in their favor, provided they pray diligently and repent their sins.

— The averagely informed person needs self-defining terminology. Family-tree heredity is one such term. Sporadic or one-shot heredity is another. The teacher may explain the difference as follows:

Sporadic heredity, as its name implies, is not expected to occur more than once, though there is no absolute guarantee that such will be the case. Though sporadic heredity can be explained in lay terms without reference to chromosomes and genes, it is preferable to give a simple lesson on cytogenetics. Pictures are indispensible—a photograph of a cell (Figure 7-1) of a spread of chromosomes and their arrangement in a karyotype (Figure 7-2) and a simplified diagram of genes. One great advantage of such pictures is that they help people to depersonalize the responsibility for heredity, and to relieve themselves of personal blame, guilt, and responsibility. Pictures also give the mind an image on which to attach explanations of sporadic changes in the heredity of egg, sperm, or zygote as a product of spontaneous mutation, or of interference by means of a drug or radiation effect, viral invasion, or by reason of the loss, gain, or translocation of one or more chromosomes or part thereof. (For karyotypes of some of the more common sex chromosome abnormalities see Figures 7-3 to 7-7.)

Family-tree heredity may be transmitted by a parent who is a hidden carrier. That is to say, he or she does not have the trait or condition himself or herself as an open carrier, but carries the hereditary package hidden away in the genes. Hidden carriers may look back in the family tree and find no known open carriers. This happens especially in that kind of hidden carrying (Mendelian recessive) in which a hidden carrier may be able to pass on an open condition only if, by gambler's luck, he or she happens to be matched with a partner who also is a hidden carrier. Then, each time the couple conceive a pregnancy there is one chance in four that both the egg from the mother and the sperm from the father will be hidden carriers. The result is that the baby will be an open carrier and will manifest the trait or condition. Likewise, there is a one-in-four chance that both the egg and the sperm will not be hidden carriers. Then the baby will not be a carrier of the trait or condition. If only the egg or the sperm, but not both of them, should be a hidden carrier, then the baby will, like the parents, be a hidden carrier of the trait or condition, but not showing it—and the chances are two in four that this will happen when each parent is a hidden carrier.

A baby who is a hidden carrier will be able in adulthood to have children who are open carriers only if he or she mates with another hidden carrier. This is the principle that lies behind the warning against the marriage of close relatives. There are more chances for one hidden carrier to match with another if both come from the same family tree than if they are unrelated. That is undesirable when both are hidden carriers for an undesirable trait or condition, like a disease; but it is desirable when the trait or condition is a good one, like superior ability. For this reason, animal breeders build up superior stock by interbreeding parents with their own offspring and brothers with sisters.

There are some traits or conditions that can be transmitted to an offspring when only one parent is a hidden carrier. In this circumstance the parent who is the hidden carrier belongs to one sex, and the child who is the open carrier belongs to the other sex (sex-linked dominant heredity). For example, some queens and princesses of Victoria's line were hidden carriers for hemophilia (bleeder's disease), whereas they had sons who were open carriers.

A person who is an open carrier of a trait or condition may transmit it to an offspring regardless of whether the other parent is a carrier or not and regardless of the sex of the baby. The trait or condition is then identified as a genetic dominant. This is the type of heredity that most people think of when they say that a child takes after his father, uncle, grandfather, great grandmother, or some other person in the family tree.

The ideal in genetic education is that it is given on a personalized tutorial basis. Economy of time, however, requires that illustrated teaching booklets and movies, still to be produced, be utilized in addition to tutorial discussion and retrieval of knowledge gained.

Counseling: Prophecy versus Probability Prediction

Medical experts have a long history of making prophecies about their patients' health and likelihood of death. Patients have an equally long history of longing for and fearing such prophecies, and of delighting in proving them wrong. It is a cardinal principal in educational genetics that one never makes prophecies, but only actuarial or probability predictions.

Probability or chance can be either exciting or dangerous to live with.

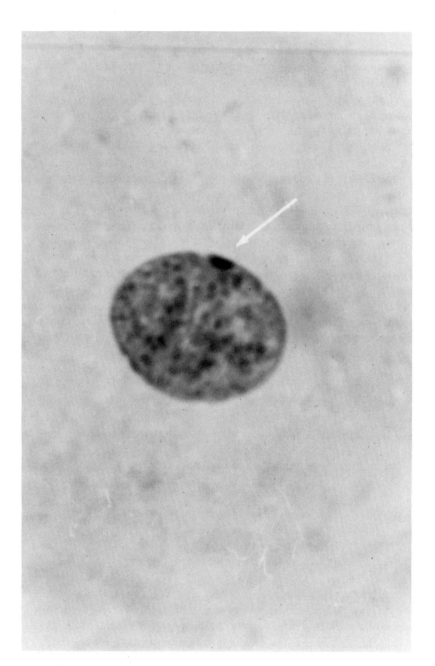

Figure 7-1. Nucleus of Cell Showing (Arrow) Sex Chromatin or Barr Body

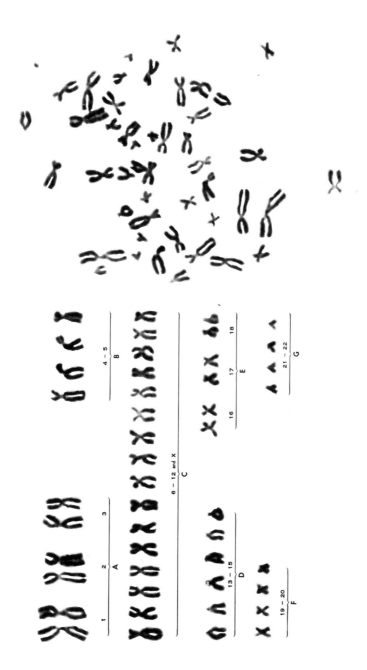

Figure 7-2. Karyotype of the Chromosomes (46,XX) of a Normal Female

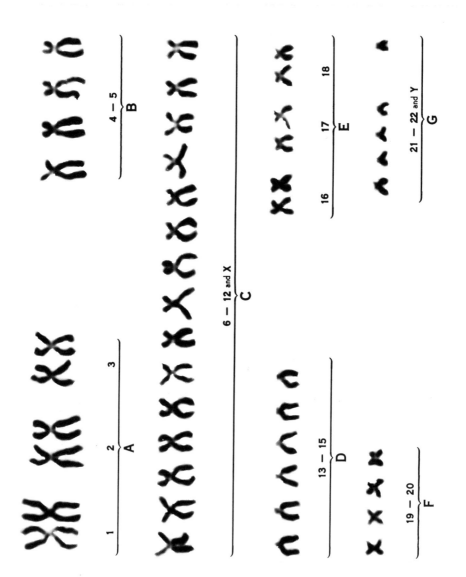

Figure 7-3. Karyotype of the Chromosomes (46,XY) of a Normal Male

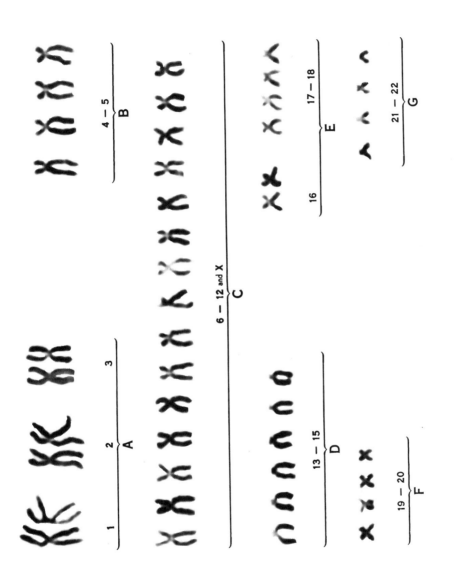

Figure 7-4. Karyotype of the Chromosomes (45,X) of a Morphologic Female with Turner's Syndrome

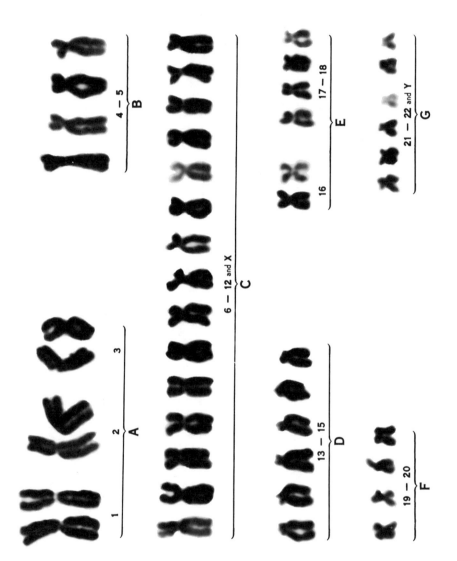

Figure 7-5. Karyotype of the Chromosomes (47,XYY) of a Morphologic Male with the XYY Syndrome

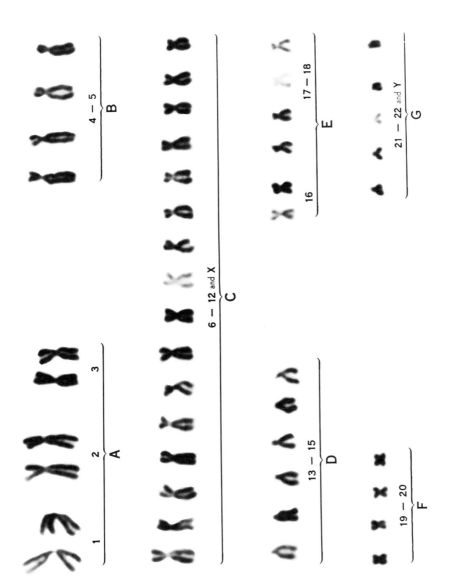

Figure 7-6. Karyotype of the Chromosomes (47,XXY) of a Morphologic Male with Klinefelter's Syndrome

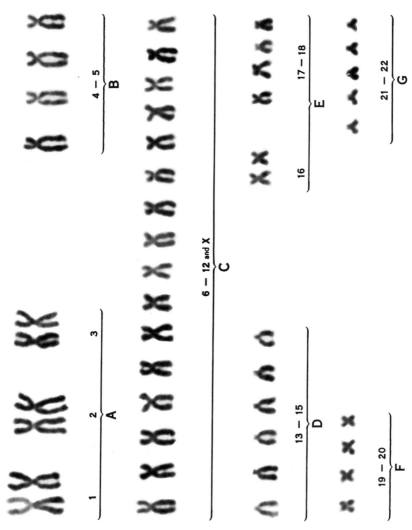

Figure 7-7. Karyotype of the Chromosomes (46,XX, Trisomy 21) of a Morphologic Female with Down's Syndrome (Mongolism)

Some people turn to Las Vegas, others turn to Hartford. For people with a gambling personality, chance and the gambler's luck is exciting. For people with an "insuring personality" it is dangerous—they buy insurance as a protection against the one-in-ten-thousand chance of bad luck. Gamblers run their risks the other way, always optimistic that good luck is so frequent that next time could not possibly bring bad luck. They fly without travelers' insurance, and forever postpone making a will, lest it be a bad omen.

People are not always thoroughly consistent, so that a person may be cautious against bad luck in most situations, but reckless and risking good luck in a few others. Since it is impossible to make an exact genetic prophecy, a couple who are at risk for a genetic defect are forced into a probability choice—either to play safe and risk not having a normal child, or to gamble and risk having an abnormal one.

Pinned on the horns of a dilemma, a person might find any promise of escape acceptable. Some formulate their escape in terms of the supernatural, looking for guidance from the occult or from religion. Some accept the dictate of authority and require their geneticist to decide for them —which pleases the physician accustomed to patient obedience, but also makes him bear a burden of responsibility that may backfire, as when a paranoid patient with a grudge kills a doctor. Some resign themselves to the dictates of genetic fate, which they may regard as an expression of the mystery of God's will, and accept both normal and abnormal children as they go ahead with more pregnancies. Some few are in the new position of being able, through amniocentesis, to predict early in pregnancy whether their baby will be normal or not, and to elect an abortion if it is abnormal —an election that requires them to have made their own ethical decision about abortion.

Part of the dilemma in deciding whether or not to get pregnant, or whether or not to abort a defective fetus, depends on how well the carriers of a genetic defect are able to foresee the long-term outcome of having a defective child. Here they must rely heavily on their genetic counselor, for they do not have the necessary data on which to make a prognosis. The prognosis needs to be threefold: the effect of the defect on the affected child; the effect of the defective child on the lives of the parents and their relationship together; and the effect of the defective child on the development of other children in the family. That's quite a challenge for any genetic counselor!

At the present time there are some genetically transmitted diseases, like Tay-Sachs disease, in which the outcome is so uniformly catastrophic and heart-breaking that it seems easy, at least for medical people, to decide that no one should wittingly give birth to an affected child. Genetic statistics are seldom so clear-cut, however. In many genetically transmitted disorders the degree of severity may vary so that the outlook is not always hopeless.

Some disorders are responsive, or partially responsive to treatment—and there is always the possibility of new therapeutic discoveries. Some disorders are late in manifesting themselves, so that the individual lives normally for part of a lifetime.

There is no consolidated source of information today to which any and every genetic educator and counselor can turn in order to obtain accurate and authentic prognostic statistics on all the known genetic disorders. Advice to parents and prospective parents must be given, therefore, on the highly variable basis of each counselor's clinical knowledge and impression. In some cases the risk is no more frightening than that of driving on the superhighway to keep a hospital appointment. In some it is less, in some greater.

The unsatisfactory state of knowledge regarding individual prognosis in cases affected by genetic defect is even more unsatisfactory with respect to prognosis of the effect of a genetically disordered child on the parents and other members of the family. Certainly, there is great variability among families, but I know of no attempt to codify the differences in such a way as to lay the basis for prognosis. Thus, today's counselors may be reduced to the assumption that other people's reactions would be the same as their own, if they were standing in the others' shoes. The assumption may well be wrong. For example, a genetic counselor, being a person with an extra-domestic career, probably would not want to sacrifice this career to the care of a defective child. By contrast, a couple whose capacity for an erotic relationship is neurotically impoverished may find that self-martyrdom with long-term devotion to the care of a defective child is the only way for each of them to maintain a relationship and rear a family.

Counseling: Recreational versus Procreational Sex

Statistical genetics has developed in the era of birth control. Without effective contraception genetic counseling could be tantamount to counseling for celibacy. For individuals who have religious, moral, or esthetic objections to the use of contraceptives, genetic counseling may, indeed, signify a recommendation not simply against breeding, but against sexual intercourse, since such persons prohibit themselves sex for pure recreation. Thus, a genetic counselor has the obligation to broach the issue of recreational versus procreational sex in the erotic relationship of a couple—whether they are married or otherwise.

The birth control era notwithstanding, this is still a touchy and sensitive issue for many couples, and one that may profoundly influence their judgment on family limitation for genetic reasons. It may require that the genetic counselor allocate time for adult sex education and counseling in the sexual and erotic relationship. There are very few people equipped by training and experience to do both genetic and sexual counseling. Sex counseling itself is a new and undeveloped branch of medicine. It is variably included in the training of marriage counselors. An effective collaboration might, therefore, be established between a genetic and a marriage counselor. Ultimately, however, genetic counselors should become qualified to some degree as sex counselors, capable of dealing with the inhibitions, anxieties, and sexual dysfunctions that may prevent or be attendant on family limitation for genetic reasons. Loss of erotic responsivity may be one such dysfunction in either the male or the female. Escape from the marriage into an extramarital affair may be another, or there may be separation and divorce.

Another sexual issue for the genetic counselor is that of the efficacy of contraception once the decision in its favor has been made. This issue is not unique to genetic counseling cases, but in all instances of family planning. Contraceptive failure is attributable in part to failure of the device, as for example when a woman wearing an IUD becomes pregnant. Also, it is attributable in part to the human element—forgetting to take a pill, for example, or impetuously having intercourse unprotected because supposedly at the safe period.

One of the virtues of a nearly 100 percent effective method of contraception, like the IUD, is that, being not quite 100 percent effective, it still allows the unpredictability of chance to take a hand in deciding one's life. For many women this is much more preferable than personally deciding on a total negation of one of the most extraordinary things a woman can do, namely, conceive, gestate, deliver, and nurse a new human being. Something of the same applies to a man in his pride of impregnation and fatherhood. There is more than simple pride of parenthood, here, for both the man and the woman are agents of a phyletic dictate as ancient in evolution as sexual reproduction itself. This dictate is responsible for proneness to accidental contraceptive failure, as well as for overt risk taking to tempt fate.

The dictate is not universal nor inescapable, however, for some people do reach a decision to become permanently sterile, namely by vasectomy, tubal ligation, or hysterectomy. The decision is difficult enough, even for those whose family is completed at the size they want. Probably for most people fertility is a very basic element of one's body schema and self-image. Those who sign it away too readily, especially under pressure from a genetic counselor, may be damned never to live in subsequent peace with their decision. Even in fantasy they cannot build a drama of their own fertility. Those who decline to sign the operative permit may be equally damned, if they do produce another doomed child.

There is no easy way out of this double bind of being "damned if you do and damned if you don't." The easiest way to get off the hook is to line up support, pro or con, from those authorities in one's life who personify the double bind. Thus, if a young couple get support from their parents and their religious counselor, as well as from close friends, in support of the genetic counselor, they can bear the burden of their decision more comfortably. Even a unanimous contrary decision eases the burden. In that case the genetic counselor needs to take special precautions not to indicate, even covertly, that he rejects his patients because they have gone against him—all the more so if the patients have lined up other doctors to take their side against him. They may need his support more than ever, if they do have another defective baby.

The counselor can always find some macabre comfort in the fact that, in the absence of breeding among genetic carriers of a given condition, he would have no problem of etiology, prevention and cure to work on, and no chance of discovering general principles of genetic health for the benefit of all. In any case, he must be watchful not to play the role of a god offended because his advice was not taken. What he may do is embark on a program of extended counseling in which parents and religious advisers are themselves given genetic education so that they are in a position to give more enlightened advice. In fact, there is a place in a comprehensive genetics counseling clinic for genetically informed clergy, and for trained lay counselors drawn from the ranks of those obliged to make their own genetic decisions. In certain instances, a genetically informed legal opinion may also be needed. The greater the number of participants, the greater the need for all to be mutually frank and outspoken with one another, and not subject to the manipulation of patients prone to play one off against the other.

It is conceivable that a jury or committee of genetically informed experts and lay persons may one day be entrusted by law with making certain decisions about the fertility of the carriers of adverse genetic conditions. Even so it will be essential to build in sturdy safeguards of the ethical rights of patients. In the meantime the right of decision rests with the prospective parents. They have the right to make a wrong decision, as well as a correct one, regardless of the criterion of what is correct and not correct.

Counseling: Psychodynamic Appraisal

A cardinal rule in counseling husband and wife or parents and child is that each should be given a chance to talk alone as well as together. They may talk to the same counselor, or each to a different one. Each person is overtly and explicitly guaranteed medical confidentiality so that nothing each says about the other will be divulged without permission.

One can adapt to the space, schedules, and other exigencies of clinical practice, but generally it is preferable for a couple to sit together initially in a brief joint interview with both counselors present. First get the purpose and schedule of the counseling program briefly defined and understood. Agree on the proposed cost, if any. Many people are justly terrified of medical bills. Clear the air of fears that patients may have of counseling because of its possible adverse association in their minds with psychiatry: Counseling means prevention, not diagnosis or treatment in psychological medicine.

The reason for separating informants is that when they are alone they tell a different story than when each is playing to the gallery in which the other is an occupant, so to speak. In a followup joint interview, the counselor then gains fresh information of their behavior, reciprocally, as a pair. This latter is particularly important as an indicator of the state of mental health of the partnership.

People's behavior can be analyzed in terms of five variables of being and three variables of doing (Lists I and II; see Money 1957).

List I: Five Variables of Being

1. Being in groups, beginning with the child-parent group
2. Being in an ecological cycle of food, shelter, and clothing
3. Being sexually male, female, or in some way ambiguous or impaired
4. Being somatically typed in ways other than sexual, e.g., tall or short, fat or thin, colored dark or light, stigmatized or not by bodily features
5. Being sick or healthy, and finally dying

List II: Three Variables of Doing

1. Inhibition and control (self-regulatory)
2. Action and mastering
3. Thinking and having images

A genetic counselor naturally enough assumes that his patients have come to talk to him about genetics. Mentally checking List I as he listens and inquires, however, he may discover that another variable looms more ominously in their agenda—for example, being unable to get along together as a family group, being poor, being unable to function sexually, being discriminated against because of skin color, being recently operated on for cancer, or being bereaved from recent loss of a parent, and so forth. First things must come first in counseling, otherwise the agenda becomes hopelessly confused, and the counseling chaotic.

The three variables of List II can be utilized to analyze any sample of a person's behavior. One, however, will stand out as more conspicuous than the other two and may, in fact, represent the person's signature or life style in most other samples of behavior. Thus, a husband in a genetic counseling session jokes that life is too short to be worrying, and says debonnairely that he leaves all the children's medical care to his wife. He earns the living and spends his weekends perfecting his golf. As further evidence emerges, it transpires that he is a man busily active with other things that mask or inhibit concern with the serious issue of whether to dare have more children or not. Action and mastery, more than inhibition and control, are his dominant style; and his thinking is at fault not because he is uninformed of his genetics problem, but because the problem is disregarded or tuned-out, and handed over to his wife to resolve.

There are many permutations and combinations of the three variables of doing, and the manifestations thereof—too numerous to present here. But a few examples will demonstrate what the genetics counselor should become attuned to. A paradoxical manifestation of a predominance of action and mastery is seen in the person, perhaps the wife more often than the husband in a case for genetics counseling, who responds to the problem of having a genetically defective child by becoming fully engaged in the role of self-sacrificing martyr. This same person seems relatively indifferent to the possibility of having a second affected child—in fact seems almost to be positive about it. The key lies in the fact that her busy martyrdom keeps her husband at a distance. She is always too busy or tired for sex, which she has always found intrusive and repugnant to her. Sexually she is inhibited and unable to let go. For her, the thinking variable functions well, except that she has no insight into the connection between self-sacrifice for the children and evasion or rejection of her husband sexually.

An example of an excess of inhibition and control might be found in the woman who is unable to ensure no future genetically defective conceptions because she and her husband both accept their religious doctrine against contraceptives. She endures her husband's sexual approaches, but freezes and finds no pleasure in sex for herself. Instead, she is likely to go into a spasm of hyperventilation, which induces a blackout spell. The very thought that her husband might want sex gives her a blinding headache, which incapacitates her for that occasion—but she does not induce a connection between inhibitory somatic symptoms and sexual avoidance.

Another example illustrates how the first two variables of doing may appear relatively unimpeded, while thinking and imagery changes. In this case a man who has had two genetically defective children has decided on a vasectomy. He has always been something of an idealist and dreamer, with a strong interest in the occult. He has a half-dreaming, half-visionary experience as a result of which he decides to resign his managerial job in order to set up his own business of mail-order, self-help, sex-education kits for married couples who cannot afford marriage counseling. He recognizes the temporal connection between the two decisions, but cannot relate the vasectomy causally to the change in his career.

Psychotherapy

Psychodynamic appraisal can be useful in the practice of genetic education, counseling, or psychotherapy. Psychodynamic appraisal should help one to recognize those persons for whom genetic education will suffice, and it certainly should help one identify the special issues requiring attention for effective genetic counseling. It will not, however, by itself alone reveal the severity of a couple's difficulty, and whether their sessions of genetic counseling should be expanded into psychotherapy for a longer term. Provided it is known, a prior history of evidence of psychopathology should be a guide here. In the absence of such retrospective evidence, the genetic counselor will have to rely on the evidence of the present. This evidence will be that the counselor himself senses he is making no headway in transmitting genetic information. In addition the recipient will be not only unable to summarize (preferably on tape) the main points that the counselor hopes he will have assimilated, but will fail in such a way as to indicate misunderstanding and misinterpretation rather than simple omission. Ideally, every genetic counseling unit will have a genetics-trained psychotherapist as a consultant.

Applied Behavior Genetics: Parents

In some babies with a genetic defect the evidence is somatically obvious at birth for all to see. If they do not already recognize the diagnosis, the parents need only to be given the name in order to seek out reference sources that will give them extensive knowledge concerning prognosis and therapy.

The situation is different with respect to diagnoses in which the evidence of genetic defect is not overt at birth and may not become clinically symptomatic even for years. Then there is an insidious temptation on the part of experts to keep covert either their findings or their suspected significance, as in the old tradition when doctors wrote their prescriptions in Latin, and consultants communicated only with the referring physician, not with the patient.

This issue of keeping findings covert has been subject to recent debate in connection with cytogenetic screening of the newborn and other sample populations for research purposes. One school of thought has favored suppressing the findings except to use them for incidence and prevalence studies. Another school of thought favors not sequestering the cytogenetic diagnosis away from the person's medical history chart, but would do no more than send it to the family doctor. The objection to this mode of procedure is that the family doctor may be in no better a position than an uninformed layman with respect to interpreting the significance of the diagnosis (of the XYY syndrome, for instance) to the parents and, later on, to the growing patient. The geneticist as counselor has the obligation to make himself an expert in these matters, and the obligation to be clinically responsible for the people who become his patients, even if they do so by the fortuitous chance of being caught in the net of one of his screening surveys.

A further justification of this assertion is that the contents of a patient's medical chart cannot be considered secret. There are too many chances for patients or parents to overhear medical conversations, or to get access to their own charts and read them—for example, when a chart is left in an examining room when the physician is called away to the phone. The knowledge thereby obtained is fragmentary, and because of its very incompleteness may fuel the flames of anxiety and its resultant symptoms. Moreover, it is the ethical right of parents and older patients to know a diagnosis and its prognosis. If the prognosis is a bad one, they may need the benefit of not building false hope, but of having someone to whom they can turn when they need to know the realities of what may happen. When the prognosis is not fatalistic, parents and patient alike may be greatly helped by the advice and counseling that can be given pragmatically only by a well-informed and clinically experienced expert. Expertise applies to counseling in the genetics of behavior as well as to prescribing medications. The XYY syndrome, Klinefelter's (47,XXY) syndrome and Turner's (45,X) syndrome, all can be used as cases in point (Money 1975).

The responsibility of the geneticist for the welfare of patients identified in research surveys means that human genetic research cannot be done independently of the provision of clinical services. Clinical research means clinical service. Clinical service includes genetic counseling, in the traditional sense. It includes also applied behavior genetics, that is to say, counseling of parents and patients for whom a genetic (or cytogenetic) condition has a bearing on behavior and its development.

Applied Behavior Genetics: Patient

The body of knowledge that constitutes applied behavior genetics in human beings is still in its infancy. The job of coding it has hardly begun. Few people yet realize how much behavioral science, pure and applied, stands to gain from systematized knowledge of behavior typical of a syndrome, irrespective of the unique individuality of the persons who happen to be the carriers of that syndrome.

An example can be found in Turner's syndrome, in which it is common for a girl to have relative difficulty in learning that requires nonverbal (performance) IQ; and to be phlegmatic and slow to emotional arousal, which is a positive asset for one who must put up with the multiple indignity of being too short in stature, dependent on exogenous hormones for sexual maturity, infertile, and in some cases physically ugly.

In a regular school system it is easy for a girl with Turner's syndrome to be victimized because her mathematical and nonverbal failures seem unjustified relative to her verbal accomplishments. Her stolid and phlegmatic response to the adversity of failure does not help, even though her personal suffering may be great. She knows that her best is not good enough to prevent failure; yet, she meets with blame and chastisement instead of enlightened understanding. It is a great relief to her to have as a counselor an applied behavioral geneticist who has an explanation for her disability (Money 1968). Instead of blaming her he blames the educational system and takes the appropriate steps to have her abilities maximized and her disabilities minimized in subsequent educational planning and advancement.

A parallel example can be found in the 47,XXY (Klinefelter's) syndrome. Low IQ is not universal in this syndrome, but is overrepresented. So also is ambulatory psychopathology, irrespective of diagnosis. It is easier for affected boys and men to benefit from counseling if they know that they are the fortuitous victims of a handicap, and not simply the behavioral product of a faulty upbringing, or of self-induced failure.

Much the same can be said of boys and men with the 47,XYY syndrome. They are from childhood on subject to personal blame and punishment for socially bad or nonconforming behavior. One of the surest ways to gain their trust and an eventual improvement in self-regulation of behavior is by way of communicating one's understanding that they are subject to a degree of behavioral impulsiveness unfamiliar to the ordinary person. Otherwise they are in the position of a color-blind person who is punished for errors of color discrimination.

Cytogenic syndromes and other known genetic syndromes provide the surest bases of knowledge for the application of behavior genetics to counseling at the present time. The application of behavior genetic counseling beyond syndrome groups to groups identified on the basis of somatic, racial or social typology is without scientific justification today. There is no methodology subtle enough to differentiate genetic from experiential determinants of behavior in nonsyndrome groups. The special criterion of a syndrome group in behavior genetic counseling is that it is homogeneous either cytogenetically or on the basis of a known and specifiable mechanism of genetic transmission.

Summary

Genetic counseling as it is known today comprises education, counseling, and psychotherapy as related to genetic transmission. In genetic education for the layman with a personal problem in genetic transmission, visual representations can be combined with self-defining concepts like family-tree heredity versus sporadic or one-shot inheritance. Probability prediction takes the place of prophecy. Some people insure against the occurrence of a dangerous risk, whereas others gamble against its nonoccurrence; and both types are encountered in genetic counseling. Recreational minus procreational sex is anathema to some couples. They equate non-breeding as tantamount to celibacy. Guaranteed nonconception through sterilization is destructive of the self-image, if fertility is an integral component of it. A psychodynamic appraisal of a couple's behavior can be formulated in terms of five variables of being, and three variables of doing. Thereby the counselor's expectancy of what the couple might gain from genetic counseling is improved. Psychotherapy as the third phase of genetic counseling is necessary only when a couple is unable to assimilate and benefit from the initial two phases. Genetic counseling can be expanded to embrace applied behavior genetics for parents and for affected patients, even though the body of basic knowledge is still scanty.

References

Money, J. 1957. *The Psychologic Study of Man*. Springfield: Ill., Charles C. Thomas.

Money, J. 1968. *Sex Errors of the Body*. Baltimore: Johns Hopkins Press.

Money, J. 1975. Human behavior cytogenetics: Review of psychopathology in three syndromes—47,XXY; 47,XYY and 45,X. *Journal of Sex Research*. In press.

Commentary I

Sheldon C. Reed
University of Minnesota

Let me open my discussion of Dr. Money's comments on a personal note—my own person, that is. I arrived at the Dight Institute for Human Genetics on August 18, 1947 to begin my duties, and during that first morning was summoned to the Pediatrics department to give genetic advice to a family in which four out of five children had some unspecified type of eosinophilia. In the afternoon there was a request for information on first cousin marriages. The next day I was asked to come to Pediatrics again to advice a family in which five of eight children had an heredomacular degeneration. I have kept a daily log of the counseling cases ever since then. There have been some 3,000 cases, more or less, so that I am long on experience but short of several skills that would often be comforting to me to have.

The reason that three counseling cases came from the same place during my first two days at the Dight was that the resident physician in charge of peds, outpatient, one Sheldon Siegel, M.D., was interested in human genetics. When he completed his duties at pediatrics outpatient, the frequent referrals from there practically ceased. The kind of mix that any genetic counseling center receives depends, obviously, upon the sources of the referrals. Every center gets an array of counseling cases different from every other center.

I have never had a counseling case for sickle cell anemia, for the XYY syndrome or even for Klinefelter's, though I did once see the brother of a Klinefelter's person who was worried that he might have children with the anomaly. I think that I was able to demonstrate to this normal XY person that his worry was not a realistic or useful worry and that it might be a good idea to worry about something else as apparently some anxiety has a positive adaptive value. I did not suggest psychotherapy to this person, as I could not see that he needed it.

The position that Dr. Money takes that psychotherapy is the third part of genetic counseling is not a new one. It sounds good, but it seems to me to depart from the real world as it is now. Franz Kallman always talked about the necessity for psychotherapy in genetic counseling, but I could never persuade him to publish or describe any cases where he had used it. Now psychotherapy does not necessarily demand any formal psychoanalysis, but if psychotherapy means anything more elaborate than the sincere desire to assist the client with his problem due to genetics, one must assume that a psychiatrist or clinical psychologist is required.

I suspect that my concept of "education" in genetic counseling is broader than Dr. Money's and that my view of education includes some sort of rudimentary psychotherapy. I do know that the counselees feel much better after they understand what their situation is. One of the greatest services that we provide the counselees is that we listen to them. We do not put words in their mouths or give directives. One of the aspects of my counseling that is greatly appreciated is that they, usually a couple, can talk with me alone. There is no circle of people in white coats sitting around to inhibit and distract us. I do get most valuable and indispensable service from other members of my "team," but they are not present at the counseling session. The invisible members of the team vary each time depending upon what is needed such as a more specific diagnosis, a chromosome study, or a biochemical assay.

However, I am more than willing to admit that I have much to learn about genetic counseling and wish that Professor Money or anyone else would present specific cases describing how the psychotherapy was done. How was it introduced?

My major research interest has been in the area of behavior genetics, which should be emphasized more than it has been. Make no mistake, I have the greatest respect for psychiatrists and would certainly involve one in the counseling if it seemed reasonable. In the few cases where the need for a psychiatrist seemed to be present, the persons had already seen one or several of them.

Professor Money touched upon the question of "hidden carriers." It seems to be that the term "hidden carrier" is confusing and should be abandoned. A carrier is a person with a hidden gene or genes by common usage and definition.

I agree wholeheartedly with almost everything that Professor Money has to say about genetic counseling. I am not objecting to his structuring of the process and agree that psychotherapy is an important ingredient of it. My caveat is that those who make an issue of psychotherapy should be specific as to how they propose to do it, beyond the personal efforts of the genetic counselor.

I was disappointed that Professor Money did nothing in the area with which he is the acknowledged expert, that is, the sex chromosome anomalies. What does he say to the Turner's, the triploid X, the Klinefelter's, and the XYY? All these persons have sufficient intelligence to formulate questions about their conditions and I wonder what these questions are and how Dr. Money answers them. As I mentioned earlier, my closest contact with the sex chromosome anomalies was the normal brother of a Klinefelter's, so I am completely ignorant of this area of genetic counseling but am anxious to learn more about it.

Commentary II

Robert F. Murray
Howard University

In this chapter Dr. Money has discussed the behavioral aspects of genetic counseling as well as genetic counseling in developmental human behavioral genetics.

I would agree that genetic counseling should be given a broad rather than narrow definition, for its concept has evolved and expanded over the past twenty years until genetic counseling is now not only concerned with the risks of recurrence of a condition of known genetic etiology, but it now includes at least the clear communication of all the medical, psychological, social, and genetic factors related to a condition either thought to be inherited or due to a disorder of the genetic material. This must include a discussion of the prognosis of the condition, as well as the consequences of different courses of action. Not only must the emotional makeup and social and religious background of the parents of affected children or of affected individuals be taken into account, but alternative programs of prevention and/or treatment must be given careful consideration.

In addition to the educational, counseling, and psychotherapeutic aspects of counseling mentioned by Dr. Money, there must also be a definite ethical, moral, and philosophical aspect to genetic counseling. This is only part of the reason why many who are involved in modern genetic counseling believe that it should be nondirective, even though from a psychiatric standpoint this is an impossible, practical position to take.

One should clearly have in mind the goals to be achieved through counseling and there must be at least the essentials of a genetic workup before counseling can be considered adequate. An accurate diagnosis is one essential of the genetic workup, but it must be supported by a pedigree of three generations and a knowledge of the scientific literature relating to the disorder in question.

Genetic counseling is part art and part science, with the greatest part of it an art like other aspects of medical science. The scientific aspect of counseling consists of delineation of the genetic disorder, its probable mode of transmission, and the detection of other members of the family who are at risk. In the context of the scientific aspects of genetic counseling, I would take issue with certain concepts presented in Dr. Money's discussion.

Dr. Money talks about carriers of genetic disorders. In this context, he uses the terms "hidden" carriers and "open" carriers. My impression is that an open carrier is someone who manifests the disease, whereas the hidden carrier is a person who does not manifest the condition in question. The definition of a genetic carrier should be broadened from that usually used. The carrier has usually been defined as a person heterozygous for a gene determining a recessive trait (King 1968). A more comprehensive definition of a carrier is a person who has within his or her genetic material an abnormality that may or may not produce definite disease in them at some time during their lifetime, but that can be transmitted to their progeny where, under appropriate circumstances (e.g., age, genetic background, proper environment, proper mating, etc.), it may produce disease. Included in this definition is not only the heterozygous carrier of mutant genes determining autosomal or X-linked recessive traits, but also the heterozygous carrier of certain genes responsible for autosomal dominant traits, as well as chromosome abnormalities in balanced form. This would include carriers of genes determining autosomal dominant traits of late onset, such as Huntington's disease since often by the time individuals have transmitted the gene to their progeny in a majority of cases they will not yet have shown unambiguous clinical manifestations. An individual may be considered a hidden carrier if there is no readily available method of determining that he or she is heterozygous for a chromosomal or single gene abnormality. But the concept of an open carrier seems invalid since an individual manifesting the disease is not in the medical genetic sense a carrier, but is affected. This distinction is not merely semantic, but has a functional medical significance.

It is in the area of carrier detection that the power of genetic counseling to prevent genetic disease can be most dramatically demonstrated. Prevention of genetic disease to any significant degree has been demonstrated by mathematical calculations to depend on the ability to detect those high risk families before diseased persons are born and/or amniocentesis can be used to provide the information and means by which couples or individuals at risk can avoid the birth of affected individuals (Motulsky, Fraser, and Felsenstein 1971). There is some evidence that behavioral and psychiatric problems are produced in a family where there are no affected children, when a high risk for giving birth to such children is revealed to the parents. The frequency, severity, and range of expression of these problems is currently unknown. This is certainly one area where research in developmental human behavioral genetic counseling would be fruitful and is needed. The hope of the medical geneticists is that at some time in the future there will be no longer hidden carriers of serious genetic defects since we will have the technology to detect easily and accurately the carriers of these disorders and may have defined the high risk populations and/or families where genetic counseling will be most useful.

There are likely to be serious ethical and moral problems in a society where this kind of carrier detection becomes routine. One the other hand, in a society where everyone is a carrier for some potentially lethal gene mutation, (which he is anyway), and knows about it, the problems of stigmatization that can result when an individual is labeled a carrier will be nonexistent.

An essential component of precise genetic counseling is an accurate diagnosis and understanding of the mode of transmission of the disorder. With the exception of certain syndromes involving mental retardation, there is a woeful lack of understanding of the precise mode of transmission of the developmental behavioral disorders. Even in these mental retardation syndromes only a few have a clearly defined pattern of inheritance. At the moment, one of the few developmental behavioral disorders that can be readily detected by screening methods and in which preventive genetic measures are available is Tay-Sachs disease, common in Ashkenazic Jewish individuals. Screening and counseling programs for this disease are in progress in several large cities. There are no rapid, inexpensive, and reliable screening methods available for other conditions. Serious difficulties exist in either making the diagnosis of the more common hereditary behavioral disorders or in understanding the genetic contribution to them. Furthermore, chromosomal abnormalities like the XYY or the XXY states, in which there is significant evidence of personality disturbances, although they can be readily detected, are not well understood. Prospective studies, now underway, will have to be completed before we can be certain of the risk of serious emotional disturbance in individuals with these abnormalities, especially the XYY state.

There are other issues raised by Dr. Money in his discussion of counseling that should be queried:

Is it wise to become involved in a discussion of sexual problems, particularly contraceptive aspects of sex, at the same time one is discussing recurrence risks for genetic disorders? Would this not have the effect of indirectly telling the counselee what to do? It is one thing to discuss certain aspects of sexual activity at the request, or as it is brought up by individuals being counseled and quite another to approach the subject as a counselor. Furthermore, such counseling takes very special skills which many in genetic counseling do not have.

How ethical is it for a counselor to counsel a couple regarding genetic risks without letting them know his or her views regarding these risks? It is certainly possible for a skilled counselor to subtly manipulate the persons he is counseling so that they will eventually follow the course of action that he or she wishes. If the couple knows ahead of time that a counselor is more concerned about the effects of genetic defects on the gene pool than the individual they will know better how to evaluate any genetic information the counselor provides. This is another reason for not giving direct advice.

Is it wise for parents to be given information about the potential of occurrence of a genetic abnormality in a child prior to the time that it is manifest and in a situation where they can do nothing to alter its occurrence? A case in point is that of sickle cell anemia where it is now possible to diagnose the potential for this disease shortly after birth. But, one does not know whether the first sickle cell crisis will occur within a few months or many years after the birth of the child. Might it not be wiser to wait until at least there is clinical anemia present in the child before discussing this with the parents? One might adhere to the doctrine of full disclosure of information, but perhaps information should be presented in stages according to the ability as, judged by the counselor, of parents or persons to deal with the information, or where the information is presented at a time or in a way that will not be emotionally damaging.

The counseling process is sufficiently complex that it is unlikely that there will ever develop the "compleat" counselor although this would be the ideal situation. To handle the total personality of the counselee as outlined by Dr. Money does require special psychiatric expertise and lots of time. When time and talent are taken into consideration the "team" approach to counseling appears to be the current approach of choice. It is premature to consider what impact effective genetic counseling might have on the frequency of these disorders or their manifestations in the population. The possession of mutant genes does not mean that the disorders produced by these genes need be manifest. With improved understanding of the genetic mechanisms that produce developmental behavioral disorders, it is possible that vigorous programs of euphenics, such as that used in phenylketonuria, will be able to ameliorate or prevent their manifestations. Genetic counseling, carrier detection, and advances in therapy will no doubt work together to significantly reduce the individual and societal burden of behavioral genetic disorders. But much more basic information about the genetic mechanisms contributing to the causation of behavioral genetic disorders, especially the more common ones, as well as a better understanding of the communication process and psychodynamics of counselor-counselee interaction is required before genetic counseling will have achieved its proper place in the armamentarium available to control behavioral genetic disorders.

References

King, R.C. 1968. *A dictionary of genetics*. New York: Oxford University Press.

Motulsky, A.G.; Fraser, G.R.; and Felsenstein, J. 1971. Public health and long-term genetic implications of intrauterine diagnosis and selective abortion. In *Intrauterine Diagnosis of Birth Defects. National Foundation, March of Dimes* 7: 22-32.

8

Ethical Issues in Human Behavior Genetics: Civil Rights, Informed Consent, and Ethics of Intervention

Arthur Falek
Emory University and Georgia Mental Health Institute

Introduction and Historical Review

Prior to the demonstration of the molecular basis of inheritance and the development of evidence for specific biochemical and cytogenetic mechanisms in the occurrence and transmission of particular human genetic disorders, heritable differences in man were described at the behavioral and morphologic levels. Thus, from a historical base, as well as in current view, human behavior genetics intersects both the biomedical and behavioral sciences.

Twenty-five years ago, when I entered the field of human behavior genetics, research studies were conducted without any special regard to ethical issues. The senior research investigators knew that they were ethical; they had selected their staff workers so they could attest to the fact that they were ethical; there was certainly no doubt that the academic and biomedical institutions where they worked were ethical; and surely the grant reviewers at the highest level were ethical or they would not have been awarded such prominent and responsible positions. In that era of naiveté and optimism in the United States immediately following World War II, science appeared to have the Midas touch from both the monetary and productive points of view. The horrors of the Nazi regime in Germany and the sadistic abuses by the Nazi doctors of their subjects in the name of biomedical experimentation, including sterilization and euthanasia, revealed by the Nuremberg Military Tribunal were well reported by the world news media. The response to those hideous crimes was the ten-point Nuremberg code (Ladimer 1970) expressing the judgment rendered by the Nuremberg Military Tribunal concerning human experimentation. That code, of course, was directed towards the despised and defeated Nazi regime. The research community in the United States at that time saw no need to establish legal precedents to deal with ethical issues in biomedical and psychological research, for they could not conceive of American scientists participating in such inhumane and nonscientific experimentation. The apprehension of American physicists about their involvement in the development of the atomic bomb was considered of little relevance to most of those in human behavior research.

In the period following World War II there were only a handful of investigators in human behavioral genetics, and they in large measure identified with the discipline of human genetics. These scientists had become thoroughly disillusioned with the simplistic notions based on assumptions of racial purity that caused the deterioration of the eugenics movement in the early part of the twentieth century in both England and the United States. They had seen the reprehensible outcome of the canonization of these erroneous beliefs in German law.

In contrast to the neglect of sound scientific method that occurred in the eugenics movement, workers in the field of human genetics painstakingly developed methods of genetic analysis in studies with a variety of animal and plant organisms. Evidence from both family and twin studies of the importance of hereditary factors in a series of defined human mental and physical disorders, and the elegant demonstration of the molecular basis of inheritance in a specific biochemical genetic disorder (sickle cell anemia) confirmed the appropriateness of these methods. Therefore, while acutely aware of the shortcomings of the eugenics movement, these investigators saw themselves as individuals whose education, experience, knowledge, and scientific methodology separated them qualitatively from those earlier workers. There was no consideration of adopting a code of ethics for human geneticists, since the medically and psychologically trained members saw their respective parent organizations as providing sufficient guidelines.

As most recently reported by Romano (1974), the basis for regulations with regard to ethical principles in medical practice and human experimentation evolved from the ancient Code of Hammurabi and the Oath of Hippocrates. The significant contribution in recent generations to the development of current codes was that presented by the Englishman Thomas Purcival, in his code of medical ethics reported in 1803. The code of Purcival was the basis for the regulations adopted by the American Medical Association in 1847, and it was the latter code that in essence was revised in 1946 to deal with the ethical issues in investigations using human subjects (American Medical Association 1958). The updating of the American Medical Association code at that time was based on the findings of medical war crimes committed by the Germans in World War II. The new code included the need to (1) obtain voluntary consent of the research subject; (2) investigate new procedures by animal experimentation prior to human experimentation; and (3) perform clinical research under proper protection and management.

A further elaboration of guiding principles for biomedical research with human subjects was incorporated in the recommendations of the Declaration of Helsinki adopted by the 18th World Medical Assembly in 1964 (Ladimer 1970). This most recent international declaration defined more precisely the basis for informed consent, and recognized the distinction between clinical research whose aim is essentially therapeutic and clinical research where the essential objective is purely scientific and without therapeutic value to the person subject to the investigation.

The first step in the development of a code of ethics in psychology was the organization of a special committee in 1938 to consider the drafting of such a code (American Psychological Association 1967). Although no code was drafted in that year, in 1939 a standing committee was appointed to deal with complaints of unethical conduct including the relationship of the psychological experimenter to his research subjects as well as to his clients and nonpsychological professionals who came to him for advice or participation in investigations. The first code of ethics for psychologists was presented in 1953, after a four-year period of development. In 1959 a series of more general principles was abstracted, and in 1963 a code of nineteen principles was established with both a judicial and an educative function. In 1973 a new ten-point code on ethical standards in psychological research was distilled from the 1963 code (American Psychological Association 1973). Like the most recent medical codes it also dealt with informed consent, the welfare of the research subject, and the ethical aspects of the research procedures to be employed.

In addition to the formulation of ethical standards by professional associations, public agencies have also become involved in the development of requirements for the conduct of research. As noted by Curran (1969), while there was continuous marked increase in funds for medical and scientific research in the United States after World War II, there were few statutes at the federal or state levels prior to the early 1960s. The two major federal agencies involved in medical research, the Federal Drug Administration (FDA) and the National Institutes of Health (NIH), permitted research investigators the freedom to pursue their research objectives guided by their own professional judgments and their own ethical standards. However, the Kefauver Senate Sub-Committee hearings of 1959 revealed the limited regulations required in clinical drug trials by the FDA and the advertisement of questionable drugs. Furthermore, the epidemic of limb deformities among newborns in western Europe due to the drug thalidomide in 1961 and 1962 initiated support for stringent FDA regulations that resulted in the Drug Act of 1962. That act fundamentally changed the laws on federal regulations required of the drug industry and those investigators working with various companies in that industry in their evaluation of new drugs. By 1966 the FDA regulations required patients' and subjects' consent based on both the Nuremberg Code and the Helsinki Declaration. At the same time the largest federal research agency, the

National Institutes of Health, was also developing ethical guidelines for human investigations.

In addition to the thalidomide disaster in western Europe, problems closer to home drew increased attention to certain abuses in human experimentation. In New York City live cancer cells were injected into geriatric patients without their informed consent. In Houston, Doctors DeBakey and Cooley at Baylor University engaged in an unseemly dispute over their priority rights in artificial heart transplant programs. Recommended guidelines with regard to moral and ethical issues in biomedical research investigations were adopted by the National Institues of Health in 1965. By 1971 the FDA added the requirement of a peer review for all clinical research, while at the National Institutes of Health authority to establish uniform policies for the protection of human subjects involved in research programs was elevated to the Department of Health, Education and Welfare. As of 1971 those policies were designed to cover all biomedical, psychological, sociological, and educational research programs using human subjects (HEW guidelines). According to the Curran report (1973), the direction in the United States at present is toward a national monolythic centralized program to deal with all legal and ethical issues in human experimentation. The report warns that such a federal centralized program is vulnerable to periodic restructuring and reorganization as a result of changing political and administrative views. The current trend, however, is towards tight controls that are being developed with little input from the scientific community. At the institutional level the regulatory systems initially designed as separate entities by the FDA and the NIH have now been essentially merged and are the function of a single review committee. This human rights committee is now evaluating not only federally funded programs, but all research projects conducted in the institution. Investigators and their institutions are told that the inferred purpose of this peer review policy is to protect them from any new criticisms and additional restrictions. However, the geometric increase in rules and regulations emanating at the federal level for the past decade belies such assurances.

The basis for court involvement in ethical issues in human experimentation, which include investigations in human behavior genetics, are derived from the new legal interpretations regarding the civil rights and informed consent requirements of all research subjects and patients. A review of some essential issues in civil rights and informed consent legislation is, therefore, presented in order to evaluate their impact on research studies in human behavior genetics.

Civil Rights and Informed Consent

The *civil rights* of patients were established somewhat more than two centuries ago by the dictum that the physician experimented at his own peril (Freund 1970). In the few court decisions in the intervening years, dealing specifically with experimentation, the physician was responsible if the novel idiosyncratic therapy he introduced was not successful where a traditional one existed. With the scientific safeguards and procedures available today, legal opinion is that the court will be hospitable to those methods that recognize the social value of human experimentation without sacrificing the interest of patients and subjects. It is suggested that this is so, since society places a high premium on scientific experimentation and judges are sensitive to community mores. The size of the concerned community varies according to the severity of the transgression and the numbers of people involved. While there was an international reaction to the German atrocities of World War II, Ladimer and Newman (1963) observe that most legal problems that arise in connection with clinical investigation do not have either international or criminal implications. Concern is with rights under civil and common law. As noted by Capron (1973), the legal rights of research subjects are based fundamentally in Anglo-American common law: the right to self-determination.

In addition to the rights under common law, there are civil rights that have been outlined in the Constitution and tested and defined in the courts. Advances in the development of an overall regulatory control in the use of human subjects in research has entailed an increase in the number of federal regulations. At the same time there have been advances in the civil rights movement evident in our political life and in our courts. The thirteenth and fourteenth Amendments to the United States Constitution; known as the Civil Rights Amendments, were enacted in the 1860s. While Section I of Article 13 proclaims the abolition of slavery and involuntary servitude, except as punishment for a duly convicted crime, the main thrust in the expansion of the rights of individuals is based on the first section of Article 14, which protects the rights of individuals and affords all equal protection of the law. The Curran report (1973) indicates that the Supreme Court, under Chief Justice Warren in the 1950s, concentrated first on the problems of racial discrimination, then on the rights of criminal defendants, and then on noncriminal defendants including the mentally ill, the mentally retarded, the alcoholic, and the drug addict. Significant precedents were established with respect to hospital commitment and care including right to treatment for all.

Informed consent according to Capron (1973) is a recognized legal right composed of two aspects—information and decision. Until recently the medical profession has not had to worry greatly over the formal details of the consent document. Current requirements were only initiated at the end of the 1950s with the expansion of the legal doctrine and the increased concern for subjects in medical research experiments (Robitscher 1974). In fact, in a review of clinical research studies Beecher (1966) indicated that as late as 1966 very few formal consent forms were used for research involving human subjects. He observed that consent forms were meaningless unless one knew how fully the patient was informed about the risks involved, and concluded that a far more dependable safeguard of the patient's rights was the presence of a truly responsible investigator.

From a historical point of view the self-determination aspects of informed consent is of ancient pedigree, while the information component is of more recent origin (Capron 1973). The joining of these two principles are the basis of the informed consent requirements in the HEW guidelines. These guidelines, it should be noted, are written to protect the research subject.

In the therapeutic setting the HEW guidelines still allow the investigator some discretion in relating information to the patient about his own illness or prognosis, although the courts most recently have called for a tightening of the language and have changed "informed consent" to "duty to disclose."

In nontherapeutic investigations the HEW guidelines clearly indicate that all research subjects are entitled to a full and frank disclosure of all information (facts, probabilities, and opinions), which an informed and fair-minded person would need before giving his judgment. The Curran report (1973) focuses on areas of vagueness in the HEW guidelines statement. The report emphasizes that it is the duty of the investigator to disclose to the subject the (1) general nature of the project; (2) kinds of procedures to be used, and (3) right of release for the subject from the project at his own request. On the other hand, they note that it is not required to inform the research subjects of (1) all the underlying hypotheses, (2) lines of investigations; (3) purposes; or (4) orientation of the project.

Parental or guardian consent is of necessity for studies involving children. To protect the investigator an attempt should be made to get both parents to sign the consent form. Even when the parents of subjects are mentally ill or retarded, but considered legally capable to perform civil functions and legal acts, they should be informed of the research program, and consent for studies dealing with their children should be obtained from them. Where one of the parents is mentally ill or retarded, the consent of the other parent would be of importance to insure the validity of the right to involve the children in the research program.

In summarizing the conclusions of Curran and Beecher (1969), as well as those in the Curran report for children age fourteen and over, informed consent is necessary from both the parents and the child in order that he may be included in the research investigation. Children under age fourteen may also participate in research investigations not for their own benefit if the studies are considered sound and hold the promise of new knowledge. It is important, of course, to assure than no discernable risks are known in the procedures to be carried out. This also applies to the parents of research subjects participating in studies at the newborn level.

If, however, an evaluation of a fetus or newborn indicates that the quality of survival will be poor, J. F. Fletcher (1966) and Shaw (1973) are of the opinion that the right of such a fetus or newborn to live should be balanced against the rights and needs of its caretakers. Both of these authors believe, however, that it is imperative that the parents must participate in any decision about treatment, and need to be fully informed of the consequences of consenting and withholding consent. O'Donnell (1974) considers such parental consent to be of a vicarious nature since the infant obviously cannot also give his consent. He considers that such consent is similar to that obtained for the markedly retarded and mentally incompetent.

For studies involving diagnosis after amniocentesis, information about the possibility of abortion as well as the purpose of the amniocentesis should be discussed fully prior to amniocentesis. If the diagnosis results in evidence of a defective fetus, then decision about abortion should be reevaluated on the basis of the diagnostic information with both parents participating in the decision regarding abortion.

Informed consent is an important issue in studies involving volunteers. Ayd (1972) in his detailed review presented evidence that about 50 percent of civilian volunteers showed psychological difficulties on further evaluation. He also indicates that those in institutions may be readily coerced consciously or unconsciously to participate in medical research. Problems of coercion or subtle pressure in recruitment of subjects come up not only with prisoners or institutionalized patients but with students as well.

The current civil rights and informed consent requirements as they relate to human investigations are under attack by consumer groups and community advocates. They are concerned with the manner in which teaching and research hospitals obtain consent from charity and ward patients. They are also concerned with the development of unfavorable social and psychological data pertaining to particular racial and minority groups.

In addition to the civil rights issues discussed above, there are related matters of confidentiality and rights of privacy. For research purposes there is the need for access to private and public records including those from schools, hospitals, and sources of vital statistics. Statistical information from schools, hospitals, and federal and state agencies is readily available, if the request is made from an agency or individual considered to be legitimate to receive the data and if the organization supplying the information is of the opinion that they will not thereby be shown in a poor light. To secure an individual patient or student record, on the other hand, it is necessary at present to obtain a release form from the subject under investigation or from his representative. The institution to whom the request is made is concerned, of course, with preserving the confidentiality of the records in their trust. Records of patients within a public mental health or retardation facility are usually most accessible to research agencies. In the past few years, however, concern about the confidentiality of records has limited their transmission without consent forms to intrastate agency exchanges. The most formidable precautions are those taken by the Veterans Administration and the Armed Forces.

Difficulties in the acquisition of records for research purposes are reflected in the types of data utilized in journal articles and textbooks. A review of recent publications in the field of human behavior genetics indicates that, in the event of a report on an individual subject, great care is employed to mask specific identification. Most often population data is published in statistical form without the presentation of identifying information about particular individuals.

The civil rights and informed consent issues of mass screening programs for genetic disorders have been reviewed in a special article from the Institute of Society, Ethics and the Life Sciences (Lappe, Gustafson, and Roblin 1972). Screening, according to the article, is a form of human experimentation, and the risks of possible psychological or social injury require that all studies be conducted in accordance with HEW guidelines to protect the research subjects. If the condition being tested for is neither amenable to treatment nor presents evidence of contagion, it is recommended that the testing programs be conducted on a voluntary basis. The basis for informed consent is discernment by the investigator that the subject has (1) knowledge with regard to the confidentiality of the information to be obtained, (2) information about procedures to be used for diagnosis and the risks involved, (3) assurance with regard to the privacy of the data and the prevention of access to it by unauthorized individuals and agencies, (4) ability to withdraw from the study on request, and (5) counseling available if requested. Indications for full disclosure in all circumstances are tempered by the possibility that such screening programs may reveal information that could psychologically or socially damage identified individuals particularly in those circumstances where information about the findings or the behavioral consequences associated with them are inconclusive.

One example is that of persons with an XYY karyotype. Although the constancy of the behavioral characteristics described for such individuals is still under investigation, the courts and other legal authorities continue to pressure for some conclusive statements from the scientific community and discount the disagreements in the literature. In a recent issue of the *Vanderbilt Law Review* a report was presented (Chandler and Rose 1973) detailing the constitutional protections and limitations available to persons concerned about their possible identification as an XYY through mass screening studies. The unjustified implication of the authors was that such individuals had an increased probability of having a violent asocial behavior pattern, and that such identification could possibly lead to the incarceration of individuals at such high risk for criminal behavior. The authors evaluated the various legal statutes available to test this issue in the courts.

The above mentioned law journal report, together with the Kennedy Senate Bill (S2072) and its complement in the House (HR10403) to amend the Public Health Service Act to give further protection to human subjects in biomedical and behavioral research, as well as the proposed policy by the Department of Health, Education and Welfare on the protection of human subjects, supports the contention that the legislative, administrative, and court actions to expand the rights of the individual are far from completed.

The investigator, on the other hand, is in a legally vulnerable position if he does not comply fully with the federal guideline requirements including its stipulations on informed consent. Since some aspects of the guidelines are written in relatively general terms, those sections open the door to legal interpretation. While the detection of such passages in the federal documents may stimulate the adversary juices of the legal profession, by design the increased tightening of the requirements hampers the activities of scientists. There are those, of course, who consider the limitation of human research investigations a fruitful result of the federal guidelines. At the same time there are others, and in fact some may belong to both groups, who criticize the slow pace of scientific studies and the unwillingness of investigators to generalize from limited results.

Once again the individual with an XYY karyotype exemplifies the conflicting issues in this situation. The emotional response from the public to news reports about these individuals stimulated legal interest and resulted in court actions based on limited scientific data. Before the necessary epidemiologic and behavioral studies were available, improper publicity prevented the opportunity to complete the essential research programs. Fear that the information obtained would be used against those identified with an XYY karyotype was not dispelled by guarantees of procedures to protect the privacy and the confidentiality of the results (Borgaonkar 1972). The observed association between those with the XYY karyotype and behavioral deviancy of a criminal nature was accepted as conclusive by the public and legal profession, while the behavioral scientists were still in the initial phases of their investigations with regard to this relationship. As indicated previously, arguments to test the legality of mass screening studies to detect such individuals based on the implications that such identification would lead to social stigma and incarceration have been reviewed (Chandler and Rose 1973).

Long term follow-up studies to obtain developmental histories of individuals with an XYY chromosome constitution and no evidence of behavioral difficulties need to be done. There is information that such people are in the population. The long term follow-up of newborns with chromosome anomalies would seem to be most essential. However, there is serious concern about whether to inform parents that their child has such a chromosome anomaly. If such information is required to be given to the parents, and it might well be based on new guidelines, would the behavioral outcome match expectancy or counteract it?

The mass media, for some reason, has not as yet seized upon the fact of the increased frequency of XXY (Klinefelter's syndrome) among persons in mental and custodial institutions. Apparently this is the reason that there is no public outcry or criminal judgments specifically directed towards this group of individuals.

The need of confidentiality and the privacy of records are major issues in such a disorder as Huntington's disease. While family history data over generations are essential for diagnosis, there is much concern that unauthorized disclosure would seriously limit training, employment, and marital opportunities. A review of these issues in genetic counseling with families with Huntington's disease (Falek 1973) has recently been reported.

With regard to studies of those at high risk of mental illness, the Curran report (1973) presents methods to comply with the letter of the law and at the same time presents a rationale to evade full disclosure that is the intent of the law. This most useful legal document contains pragmatic advice for scientists in human behavioral genetics, for it clearly spells out those legal niceties that permit the scientist to continue his investigations within the current boundaries of the law. The lesson to be learned by the scientists is that they should be aware that their inferences may be interpreted by others as facts. Biomedical and behavioral scientists need to be protected from the misuse or misinterpretation of preliminary reports in their professional journals. Perhaps informed consent of the authors should be required before such data are employed for other than research purposes.

Human behavior geneticists should learn from past experience not to voice premature conclusions outside a research setting. No matter how coercive the requests to do otherwise may be, they should be resisted because preliminary statements may later be used to discredit the profession.

Genetic Intervention

Genetic intervention is at two levels. At one level intervention is directed toward the control or prevention of reproduction. At the other, that of genetic engineering, intervention is concerned with methods to correct or modify genetic defects. A number of books and published symposia (Bergsma et al. 1972; Bergsma and Motulsky 1971; Falek 1973; Frankel 1973; Hamilton 1972; Harris 1972; Hilton et al. 1973; Robitscher 1973; Siekevitz 1972) have reported the views of scholars in genetics, law, philosophy, and religion with regard to their opinions about genetic intervention. Furthermore, the Institute of Society, Ethics and the Life Sciences has been established to focus on this and other problems in biomedical ethics.

In addition to the ethical issues of intervention raised in the many previous reviews, the human behavior geneticist's difficulties are confounded by specific involvement in the socially least acceptable types of disorder—those of mental retardation and mental illness. Physical disorders with no observable deformities evoke society's sympathy, those with visible facial or bodily defects, less so, but those that result in mental impairment (illness or retardation) society regards with distaste and rejects. Society's method of rejection is institutionalization. The human behavior geneticist should be aware that his research interests are those that often evoke emotional responses from the public, and that discussions about intervention, therefore, need to be couched in careful terms.

The most acceptable method of intervention is, of course, the one that corrects the defect in infancy before there are observable signs of the disorder. An example of such intervention, a method of genetic engineering, is that to prevent phenylketonuria. From the point of view of the individual and his family, the development of the technique to treat the individual from birth through childhood by dietary means to prevent mental retardation and lifetime institutionalization is a positive approach. From a population genetics point of view, however, there is concern that if all affected persons with this genetic disorder could be treated and reached reproductive age in good health, instead of a loss of their mutant genes from the gene pool, such individuals would survive and transmit their mutant genes to their offspring. An increase in the frequency of phenotypically unaffected individuals in the population who are carriers of the mutant gene would in the long run increase the number of affected persons. Since management for the disorder is known and available, the increase in affected individuals is now a problem in medical economics.

It is expected that in time most, if not all, recessively inherited biochemical disorders will become amenable to medical control. The question to be raised is whether genetic engineering will result in our becoming a nation of cripples requiring a host of therapeutic aides. The pessimists are concerned about the size of the portion of the gross national product that will have to be spent on medical therapy.

The optimists, however, believe that natural selection will continue to select genes with favorable expressions. That is, those best adapted to the current environment, including man-made environmental changes—medical, technical, and social, will continue to reproduce with greater advantage in the population. Since in the future genetic planning will be in human hands, it is important that we recognize the ethical issues involved.

The area of genetic engineering that focuses on both social and medical methods of environmental modification to permit genetically abnormal individuals to lead normal or relatively normal lives is called euthenics. In addition to dietary control for phenylketonuria, other measures are glasses for those with myopia or astigmatism, stapecectomy to correct the conductive hearing loss in otosclerosis, drug therapy to remove excess copper in Wilson's disease, immunological prevention of erythroblastosis fetalis (Rh disease), as well as educational facilities for the blind and deaf.

Euthenic measures also include current standard treatment procedures to deal with genetically transmitted mental disorders including manic depressive psychosis and schizophrenia. In large part the remarkable increase in the out of hospital mental health treatment programs and the opportunity to maintain patients at home have been achieved as a result of advances in psychopharmacology.

Unless former patients can be maintained outside of the hospital and behave within limits tolerated by the community, they will arouse negative emotional responses that may result in a backlash. It is also to be expected that current mental health programs will produce expressions of both concern and encouragement as pessimists and optimists view future mental health needs based on the changed potential for reproduction of affected individuals.

New therapeutic tactics, of course, should be in line with Kety's assertion: "Among experiments that may be tried in man, those that are harmless are permissible and those that do good are obligatory" (Beecher 1966).

For disorders that do not respond to currently accepted treatment measures, new procedures should be permitted only on a research basis. Psychosurgery is one such example. The National Institute of Neurologic Diseases and Stroke (Goldstein 1973) after careful review has recently recommended the continuation of therapeutic interventions including surgical procedures to investigate the appropriate treatment of uncontrollable rage, but only under most careful and monitored conditions. Other treatment modalities that deserve long-term scientific evaluation include the various behavioral modification procedures now employed or under consideration for use in prisons as well as institutions for the mentally ill or mentally retarded. There are many difficulties in conducting such studies in the emotional climate of our times.

In addition to treatment for those with overt disorders there are also programs directed toward the identification of those at high risk of demonstrating a behavioral genetic defect. Some of the issues involved in such identification were dealt with in the discussion about individuals with an XYY karyotype. Intervention to identify children at high risk for schizophrenia is based on evidence of illness in one or both parents. The Curran report deals with the ethical and legal issues of such intervention, and notes that once the accuracy of identification is achieved, there will be strong pressure to develop treatment programs. It is their recommendation that workers in this area maintain a "low profile and keep out of court involvements regarding civil rights and informed consent issues." The strategy of current efforts, according to the Curran report, is to support studies of children at high risk for schizophrenia, which avoid possible harm and contain health and environmental factors to improve developmental outcome for all children.

Similar kinds of care should be taken in the development of programs that attempt to modify the environment in order to improve the level of intelligence in culturally deprived populations. The environmental approach requires an upgrading of educators and faculties in school systems as well as the development of family training programs and infant and early care centers for those born into culturally deprived households. Such programs should permit each child the opportunity to attain his highest level of achievement.

A survey of studies on genetic aspects of intelligence (McClearn and DeFries 1973) includes evidence that (1) genetic factors are important; (2) there is considerable overlap in intellectual capacity among different occupational groups; (3) many children are not intellectually characteristic of their father's occupational class; (4) assortative mating even within an occupational class is based on personal choice; and (5) heritability as a measure of intragroup comparison is at best unclear. From these findings it is apparent that arguments about the intellectual superiority of one race, religion, or ethnic group over another cannot be substantiated.

In contrast to euthenic measures to improve the intellectual level in populations, negative eugenic programs have been promoted in an attempt to limit the frequency of individuals with low intelligence through the control or prevention of reproduction. One suggested approach is through the development of family planning programs that will be accepted by reproductive individuals of low intelligence.

Institutionalization has been another method to prevent the reproduction of those with mental illness or retardation. Since most disorders are neither precisely defined nor inherited as simple autosomal dominants, this has done little to decrease the frequencies of these disorders from one generation to another.

Likewise, the proposal to establish sperm banks to collect and store the sperm of socially desirable men to be used for the fertilization of selected women as a technique for the preferential breeding of superior persons, positive eugenics, is also an unsuccessful method to control the direction of evolution.

Genetic counseling is another negative eugenic measure. While implicit in counseling is the presentation of information directed towards the limitation of reproduction in couples at high risk of having a child with a genetic defect, this approach is now modified in accordance with the ability to identify and abort the abnormal fetus or treat the child after birth.

In such late onset disorders as Huntington's disease, there are times, of course, when diagnosis places the genetic counselor on the horns of a dilemma. Present day medical ethics indicates that every patient has the right to know the results of the diagnostic evaluation. Some discretion about the dispensation of such information has to be given for those patients who show suicidal tendencies or other emotional instabilities. Counseling in which emotional issues have not been taken into account may be more hazardous than no counseling at all, and would only be likely to aggravate existing fears, anxieties, and inner tensions that by their very presence would tend to prevent acceptance and realistic interpretation of the given information. Difficulty with regard to acceptance of the diagnostic information is often observed in those cases with no phenotypic evidence of disorder. Even with a clearly visible disorder like Down's syndrome, parents often deny the diagnosis exclaiming, "God would not do this to us." The counselor should be aware that similar problems in coping may produce psychological difficulties in other family members seen in counseling sessions. The genetic counselor, therefore, should be trained to recognize the behavioral complexities associated with counseling as well as be professionally qualified in the field of genetics. Diagnostic capability is useful but is not necessarily part of the armamentarium of the genetic counselor.

Abortion is another method to control reproduction. When there is concern about genetic disorder, decision about abortion is usually based on biochemical or cytogenetic information obtained after amniocentesis. There are now more than thirty inborn errors of metabolism, and, in addition, cytogenetic anomalies that can be detected during the first trimester of pregnancy by this method. Motulsky (1972) has cautioned that case reduction of autosomal recessive disorders by abortion after amniocentesis initiated only after the birth of an affected child will be relatively small. To obtain maximum benefits for the elimination of such disorders it will be necessary to institute prereproductive mass screening carrier detection programs to identify the relatively small number of at risk matings. Murray (1972), however, emphasized that the basis for such a program is due to the fact that most hereditary disorders detected at screening are not amenable to therapy. The therapeutic aspects of amniocentesis after abortion are, therefore, for the parents and not for the affected fetus. In fact, one problem of concern is how affected would the fetus have to be to justify abortion. While most would agree to abort a fetus with Down's syndrome, how many would be of that opinion if the fetus was identified as one with an XYY, XXY (Klinefelter), or XO (Turner) syndrome. How would one present such information to parents without influencing their decision about abortion, given the incomplete nature of our present knowledge?

Voluntary sterilization appears to have become more acceptable as a method of fertility control, particularly in completed families with normal children. Involuntary sterlization, on the other hand, remains a highly charged emotional issue because it permanently alters the individual's ability to reproduce without his informed consent. There are those who continue to propose sterilization as the method to prevent the transmission of observed defects, especially mental impairment. In accordance with the medical maximum, "above all do no harm," concern is with regard to which traits should be the ones for which such action is requested. Since we are all carriers of genes that in the homozygous state can produce defects, all of us are living in glass houses readily shattered by moralistic attitudes.

For the future it is expected that methods to manage genetic disorders will include procedures to enable replacement of a missing enzyme, the stabilization of unstable enzymes, in vitro fertilization, cloning, and for the not so forseeable future, gene therapy through viral repair.

Hirschhorn (1972) commented that the two concepts which require attention in dealing with the ethical issues of intervention are those of freedom and coercion. He noted that there were three groups involved: (1) families of affected individuals; (2) research scientists and genetic counselors; and (3) government representatives. Since all have different objectives and interests, conflicts will arise as attempts are made to answer questions about freedom and coercion as they effect limitations imposed on individuals as against society. If decisive stands are not now taken by human behavior geneticists, decisions will be made without them.

References

American Medical Association. 1958. Principles of medical ethics: Opinions and reports of the judicial council. *Journal of the American Medical Association* 167. Special Edition.

American Psychological Association. 1967, *Casebook on ethical standards for psychologists*. Washington, D.C.: American Psychological Association.

_____. 1973. *Ethical principles in the conduct of research with human participants*. Washington, D.C.: American Psychological Association.

Ayd, F.J., Jr. 1972. Motivations and rewards for volunteering to be an experimental subject. *Clinical Pharmaceutical Therapists* 13: 771-81.

Beecher, H.K. 1966. Ethics and clinical research. *New England Journal of Medicine* 274: 1354-60.

Bergsma, D.; Borgaonkar, D.S.; and Shah, S.A., eds. 1972. Advances in human genetics and their impact on society. *Birth Defects: Original Article Series* 8, no. 4.

Bergsma, D., and Motulsky, A.G., eds. 1971. Symposium on intrauterine diagnosis. *Birth Defects: Original Article Series* 7, no. 5.

Borgaonkar, D.S. 1972. Recent developments in human genetics—their usefulness and impact on society. *Birth Defects: Original Article Series* 8.

Capron, A.M. 1973. Legal rights and moral rights. In *Ethical issues in human genetics,* eds. B. Hilton, D. Callahan, M. Harris, P. Condliffe and B. Berkley. New York: Plenum Press.

Chandler, J.A., and Rose, S.F. 1973. The constitutional dilemma of a person predisposed to criminal behavior. *Vanderbilt Law Review* 26: 69-103.

Curran Report. 1973. *A Review of legal precedents and medical ethics involved in intervention with children at risk for schizophrenia.* Boston: Socio-Technical Systems Associates.

Curran, W.J. 1969. Governmental regulation of the use of human subjects in medical research: The approach of two federal agencies. *Daedalus* 98: 542-94.

Curran, W.J., and Beecher, H.K. 1969. Experimentation in children. *Journal of the American Medical Association* 210: 77-83.

Falek, A. 1973. Issues and ethics in genetic counseling with Huntington's disease families. *Psychiatric Forum* 4: 51-60.

Fletcher, J.F. 1966. *Situation ethics: The new morality.* Philadelphia: Westminster Press.

Frankel, M.S. 1973. Genetic technology: Promises and problems. *Policy Studies in Science and Technology,* Monogr. #15. Washington, D.C.: The George Washington University Press.

Freund, P.A. 1970. Introduction to the issue "ethical aspects of experimentation with human subjects." *Daedalus* 98: 7-14.

Goldstein, O. 1973. *Report on the biomedical research aspects of brain and agressive violent behavior.* National Institute of Neurological Diseases and Stroke.

Hamilton, M. 1972. *The new genetics and the future of man.* Grand Rapids: W.B. Eerdmans Publishing Co.

Harris, M. 1972. *Early diagnosis of human defects: Scientific and ethical considerations.* HEW Publication (NIH) 72-25. Washington D.C.: U.S. Government Printing Office.

Hilton, B.; Callahan, D.; Harris, M.; Condliffe, P.; and Berkeley, B., eds. 1973. *Ethical issues in human genetics.* New York: Plenum Press.

Hirschhorn, K. 1972. Practical and ethical issues in human genetics. *Birth Defects: Original Article Series* 8, no. 4.

Ladimer, I., ed. 1970. New dimensions in legal and ethical concepts for human research. *Annals of the New York Academy of Science* 169: 590-93.

Ladimer, I., and Newman, R.W. 1963, Legal review and analysis: Introduction. In *Clinical investigations in medicine: Legal, ethical and moral aspects*, eds. I. Ladimer and R.W. Newman. Boston: Law-Medicine Research Institute.

Lappe, M.; Gustafson, J.M.; and Roblin, R. 1972. Ethical and social issues in screening for genetic disease. *New England Journal of Medicine* 286: 1129-32.

McClearn, G.E., and DeFries, J.C. 1973. *Introduction to behavioral genetics.* San Francisco: W.H. Freeman.

Motulsky, A.G. 1972. Genetic therapy: A clinical geneticist's response. In *The new genetics and the future of man*, ed. M. Hamilton, Grand Rapids: W.B. Eerdmans Publishing Co.

Murray, R. 1972. Screening: A practitioner's view. In *Ethical Issues in Human Genetics*, eds. B. Hilton, D. Callahan, M. Harris, D. Condliffe, and B. Berkley. New York: Plenum Press.

O'Donnell, T.J. 1974. Informed consent. *Journal of the American Medical Association* 227: 73.

Robitscher, J. 1973. *Eugenic sterilization.* Springfield, Ill.: Charles C. Thomas.

Robitscher, J. 1974. Informed consent. In *Social-legal uses of Forensic Psychiatry: Teaching packet.* Personal communication.

Romano, J. 1974. Reflections on informed consent. *Archives of General Psychiatry* 30: 129-35.

Shaw, A. 1973. Dilemmas of "informed consent" in children. *New England Journal of Medicine* 289: 885-90.

Siekevitz, P., ed. 1972. The social responsibility of scientists. *Annals of the New York Academy of Science* 196, Article 4.

Commentary I

Bruce K. Eckland
University of North Carolina

Dr. Falek has presented in this chapter an excellent overview of some of the central ethical issues in human genetics, cast quite appropriately in a historical context. I will begin on a rather specific point regarding genetic intervention that he raises and then quickly move on to one or two more general questions.

Although preventing the reproduction of those with mental illness and retardation probably is more a latent than a purposive function of institutionalization, is Falek correct in proposing that institutionalization has done little to decrease the frequency of these disorders from one generation to the next? As some of you may remember, earlier investigators once predicted that intelligence might be declining in Western societies because of an apparent negative correlation between fertility (or family size) and IQ. Yet, several recent studies have shown that the correlation itself was indeed only apparent and not real. Earlier writers, among other errors, had failed to consider the very low fertility rates for severely retarded persons, many of whom are institutionalized, have no children at all, and therefore did not show up in the calculations. When childless adults are taken into account, the correlation between IQ and fertility essentially disappears. Now, if persons who are severely retarded, instead of having no children, reproduced at a rate comparable to those who are only mildly retarded (a rate incidentally that is well above the population mean), might there have been some element of truth in the earlier prophesies?

The broader question of just how intense selection pressures must be before one might observe any significant change in the frequency of mental illness or retardation, of course, is still largely unanswered and perhaps will remain so for some time. Yet, the question could be an extremely important one since a number of the more sensitive issues concerning genetic intervention generally assume that relatively significant changes in the distribution of such traits could occur as a result. This is true both of increasing the genetic load for deleterious genes resulting from improved genetic engineering, as Falek notes, and of "eugenics" programs that more directly but selectively attempt to limit or otherwise control reproductive behavior.

In regard to the latter Falek dwells mainly on negative programs, no doubt because this is where most of the activity has been. (Programs designed to selectively promote fertility have seldom been given serious attention.) People do not actually *want* to give birth to a mentally retarded or otherwise affected child and generally will take whatever preventative measures necessary if the risks are high and if given the opportunity. It is for this reason that programs of prereproductive mass screening followed by routine amniocentesis or other follow-up procedures for couples at high risk will quite likely succeed and I expect will be taken largely for granted in the not too distant future.

I am not suggesting that there are no problems to be overcome with regard to such programs, particularly if they are suspected of being coercive or are aimed at specific groups (which is exactly why blacks cry "genocide" whenever William Shockley, as well-meaning as he may be, attempts to take the platform). Nevertheless, the health benefits of preventative measures usually are recognized and accepted both by individuals, who generally must voluntarily consent, and by the society as a whole. Actually in most countries individuals are allowed a great deal of self-determination and choice in the matter on one's right to bear children. Most contemporary efforts to control man's reproductive habits are not coercive but depend greatly upon the willingness of individuals to make use of the medical knowledge and technology available to them. And, quite probably as a consequence, when the individual's interests, as he perceives them, and society's interests conflict, such efforts ordinarily are doomed to fail.

The clearest demonstration of this point that I can think of involves the high birth rates and low death rates in underdeveloped nations. There has been a natural *willingness* by most people in these countries to accept the obvious benefits of modern medical technology to save lives, which has been especially important in lowering infant mortality rates and deaths of women due to childbearing. On the other hand, there has been a general *unwillingness* to use available devices to limit population growth. Family life is still largely tied in these countries to an agrarian economy to which may be related the fact that birth control has no clear or obvious appeal for the individual. The result has been an unprecedented world population explosion of the kind Malthus certainly never dreamed of. There are two points to be made here: First, individual human values indeed *are* quite important to consider, particularly in the short run; and second, individual values under certain conditions may very substantially *conflict* with what may be in the best interests of the society. When this happens an individual's rights to self-determination may come into question and, in some cases, could be overruled, although in practice they seldom are.

Much of Falek's discussion deals with questions of individual freedom and coercion. He quite properly notes that considerable attention has been given in recent years to developing new mechanisms for safeguarding an individual's rights, primarily through informed consent. However, Falek tends to sidestep the question of what happens in regard to intervention programs when the interests of an individual, as he perceives them, and the interests of society, based on some form of consensus, are not in accord

—when the choice between freedom and control must be made. The recent history of population programs in underdeveloped nations again illustrates the primacy of the question. In spite of the very large gulf between individual and national interests, nearly all population programs stubbornly continue to stress self-determination and the voluntary use of contraceptive measures in family planning. Coercion is very seldom used in these programs. Not even government incentives are very popular. It is partly for these reasons perhaps that not many inroads have been made into the problem. With few recognizable alternatives or incentives for change, especially in regard to the traditional roles and status of women, families continue to be large in size and most governments in these countries, even if aware of how serious and self-defeating the consequences of overpopulation can be, have no enthusiasm for imposing formal sanctions or for otherwise dealing more directly with the problem.

My final comments have to do with an issue not raised in Falek's discussion but that is the central thesis of Amitai Etzioni's recent book *Genetic Fix*. That is, will we let the instrumental forces of eugenics and genetic intervention drag us blindly toward their own ends or will we establish the mechanisms and social institutions necessary to guide science and technology? Etzioni believes that the social and ethical issues raised by genetic intervention are far too great to be left solely in the hands of either scientists or politicians, and I would add, of the legal profession.

The question goes beyond that discussed by Falek on the conduct of biomedical research and the use of human subjects. It also goes beyond the general guidelines established by HEW, by university review committees, and by various professional organizations for handling those matters. What in addition is required is a mechanism for dealing with such questions as how scarce resources should be allocated in the development and dissemination of the new knowledge and techniques stemming from biomedical research, who will benefit from these developments, and what kinds of choices should people have given the known power and risks of such techniques. To find the golden mean, Etzioni calls for a permanent National Health-Ethics Commission independent of the government to monitor the standards of all biomedical research and practice, with perhaps local state and town boards, each composed of members from not only the medical profession but from theological, humanist, and other scientific communities as well.

While many of us may find fault with the idea and may think that carrying out such a proposal would lead to overregulation of our research enterprise, it would be well to consider that the probable alternatives, as Falek seems to imply, may be far worse—that is, a monolithic centralized program run by the government with a trend toward tighter controls and vulnerable to unpredicted shifts in political views.

Moreover, as I understand it, a National Health-Ethics Commission and subsidiary groups would be designed not to supplant but to complement other ethics boards and the peer review system under which we currently operate. Whether such a commission would have actual regulatory powers I presume would be a subject of considerable debate. In any case, its main function would be to routinely edit our progress and, as Etzioni states, to "make it possible to curb some undesirable effects without hindering the mainstream of development."

In closing we should not forget the cause for alarm. The deliberate use of genetic research for social ends, as Falek notes, is generally associated with fascism and the horrors of the Nazi regime in Germany; unfortunately, it still is for many of those who remember the Nuremburg trials or the racist attitudes in America in the early part of this century and that, to some extent, still pervade American life. And, yet, we are about to embark on a revolution that some believe could "do to our genes and brain chemistry what the Industrial Revolution did to our muscle power." But we do not want to be caught again on a runaway train without a mechanism for recognizing our options. We must *plan* for the conditions that would allow us to make moral decisions about our future.

Commentary II

James F. Crow
University of Wisconsin

One can only applaud the humanitarian feelings and the concern for individual rights that lie behind the recently imposed restrictions of research involving human subjects. The issues involved and the current practices have been most ably presented in this chapter by Dr. Falek. As he makes clear, we have moved rapidly from a situation where the individual investigator largely determined the manner in which his subjects were used to one where there is a great deal of regulation.

This concern for ethical issues is welcome; but there is a serious problem. Designing good research programs where human subjects are involved is always difficult, and this is especially true for behavioral studies where careful experimental design is often required to minimize biases that are frequently very subtle. The more complicated the regulations; the more red tape; and the more people that are involved in approving the procedure, the more likely it is that optimum experimental design is sacrificed. There is danger that important research with great potential for alleviation of human suffering will be hampered or rendered impossible by overly cumbersome procedures designed to protect individual rights. The problem is to find the right balance between the right of the individual subject not to be harmed or seriously inconvenienced by the research procedure and the needs of society for the knowledge that could come from this research.

Many behavioral studies, including behavioral genetics, can be done with no trauma and only minor inconvenience to the subjects. The inconvenience is often more than compensated by the personal satisfaction to the subject from participation in a research program. It will be particularly difficult to do good research if the threat of a legal suit by a disgruntled individual becomes an accepted way to halt a research project. We have too many examples in our society where a resolute minority imposes its will on a politically ineffective majority. I recently read of a suit being brought to stop a research project in the public schools designed to study the relationship between the size of the classes and the amount learned on the ground that the students were being given mental tests. While appropriate concern for the rights of the subject is needed, it will be tragic if important research of great social benefit and of promise for alleviation of human suffering is stultified by needlessly strict regulations, time-consuming procedures, excessive red tape, and the fear of legal suits.

Ethicists are increasingly involved in genetic research and heredity counseling. Their views are often derived from some general principles, sometimes of religious origin. These views may conflict with the immediate humanitarian concerns of the counselor for his client. One possibility is that amniocentesis and abortion, or artificial insemination will be withheld with the price of unnecessary human suffering. To take another example, it is an uncomfortable situation when our researchers have to go abroad for research projects because we are too ethically concerned, or perhaps only too fastidious or squeamish, to organize record linkage systems or to permit research using aborted embryos.

Somewhat similar issues arise with privacy and confidentiality. On the one hand, individuals are entitled to protection from invasion of privacy. On the other hand, the public could benefit by the results of research that entails some intrusion. It has been mentioned several times in this book (see Chapters 2 and 3) that the children of identical twins are the genetic equivalent of half sibs and that the legal uncle is a genetic father. This offers a chance to study half sib and parent-child covariances in different environments and without the possible effects of disrupted homes that are involved when separated half sibs or foster children are studied. A similar issue arises with the use of co-twin controls for study of environmental factors, such as different educational regimes or nutritional supplements. The empathy and extreme facility of communication between identical twins are also interesting and potentially important subjects for research. Many experiments of great social value could be done with most trivial harm and inconvenience to the subjects, but with some intrusion on privacy. Again, somehow we have to strike the best balance between the danger of reckless research and the frustration of overzealous protection.

In my opinion the right of the individual to determine his own behavior should be almost absolute, being limited only at the point where his activities interfere with the rights of others. But freedom to reproduce is, I think, of a different sort, for the decision to have a child also involves the rights of that child and of the population into which it is born. Recent birth rate trends in the United States are such that, except for immigration, the population is expected to stabilize once the age distribution equilibrates. But this is by no means true of the whole world. The right to reproduce at will cannot long remain a right that cannot be abridged. Infringement on the parent's right to reproduce is much less of a threat to civil liberties than condemning a child to a lifetime of near starvation. Let me also add my own ethical belief that no person has the right knowingly to produce a child with a severe physical or mental impairment. The greater right is that of the child to have a fair start.

We have heard a great deal of proper concern for the rights of the person who is the subject of research. In behavior genetics it is also necessary at this time to show some concern for the right of the researcher to do the work, to publish the results, and to teach and discuss his views. There have been instances where academic freedom and the right of free speech have been seriously abridged. We need to assert that human behavioral genetics, including the inheritance of human intelligence, is a proper subject for inquiry and for discussion.

Research Strategy in Developmental Human Behavior Genetics

K. Warner Schaie
University of Southern California

Although the theme of this book concerns the implication of genetics for developmental phenomena, scrutiny of the research data referred to suggests a curious lack of attention to the dimension of time across which such developmental phenomena must occur. Further, little attention is given to the fact that the timelines involved in changes in gene activation patterns and their consequent impact on developmental phenomena may be quite different than those involving environmental influences. Most of the data sets considered involve either cross-sectional studies (e.g., those on assortative mating and on heritability differences across populations) or single cohort longitudinal studies (e.g., most of the twin studies and animal experiments). Behavior geneticists seem to be relatively unaware of the different threats to the internal validity (Campbell and Stanley 1963) of studies using such methods, a problem that only recently has come to the full attention of developmental psychologists.

No attempt is made here to treat the full complexity of developmental design problems, but a brief summary is given indicating the limitations of currently favored data collection schemes with respect both to differences in level of function and to changes in variance components or heritability ratios occurring as a consequence of developmental events. For more detailed treatments the reader is referred to Buss (1973), Cattell (1970, 1973), Nesselroade and Baltes (1974), Schaie (1965, 1973, 1976), Schaie and Gribbin (1975), and Wohlwill (1973).

Age Changes, Age Differences, and Sociocultural Change

A good deal of recent methodological controversy in developmental psychology regarding the interpretability of research data has focused upon the often conflicting results obtained from cross-sectional and longitudinal studies, particularly upon the topic of decrement with age in the very ability

---

Preparation of this chapter was supported in part by research grant HD08458-02 from the National Institute of Child Health and Human Development. I am grateful to John DeFries, Elving Anderson, and Gerald McClearn for their helpful comments on a first draft.

measures so frequently used as dependent measures by human behavior geneticists (see Schaie 1974).

Schaie (1965) has shown that both the cross-sectional and longitudinal methods are special cases deducible from a general developmental model. The model also suggests a third possible approach, the time-lag method, in which samples of the same chronological age are compared at different points in time. Analysis of the general developmental model suggests that each of the three methods of data collection confounds a different combination of the three components of chronological age (A), time of measurement (T), and time of birth (cohort) (C). As a consequence, data from any two methods of descriptive analysis will frequently yield discrepant results and where agreement is found, such agreement may still simply reflect artefactual combinations of confounds.

A cross-sectional study at one point in time compares samples of subjects who differ both in age and cohort membership. It is basically a separate sample pretest-posttest design. That is, the older group is assumed to have had "aging" as the experimental treatment. But many other "treatments" may have occurred during the time period of environmental exposure for which the two samples do not overlap. Consequently, differences between age groups in cross-sectional studies may be attributable to age differences and/or cohort differences, while findings of no difference could be attributable to age and cohort differences operating in opposite directions. In any event it is not proper to infer age changes from cross-sectional inquiries, unless experimental isolation of research subjects has been accomplished. Indeed, the most parsimonious interpretation of differences obtained in cross-sectional studies is that of evidence for differential performance for population cohorts who differ in time of birth (and who coincidentally, of course, must also differ in chronological age). Upon replication, such differences are likely to be found not for the same age groups, but rather for the same cohorts over a different age range. For example, a difference in ability level found between forty and fifty year olds in 1975, if attributable to differences in the two cohorts' high school education, would be expected in 1985 to characterize the relation between fifty and sixty year olds. At that time there should then be no difference between forty and fifty year olds unless a further cohort difference were to have occurred.

The single cohort (traditional) longitudinal study compares the behavior of a sample of subjects belonging to the same cohort at two or more points in time. It is a time series, where once again aging is assumed to be the treatment effect. Here, age changes are indeed investigated, but they are confounded with unspecified treatment effects that are characteristic of the time of measurement during which the age change occurs. Observed differences might therefore be due to maturation or other age-related change, or to environmental events (sociocultural change) that might have similar impact on organisms at any age. In addition, the validity of such single cohort longitudinal studies is threatened by the effects of repeated measurement, experimental mortality (subject attrition) statistical regression, and their possible interactions (Schaie 1972a).

Of concern to the behavior geneticist, here, would be the problem that the phenotypic developmental change described in a single cohort longitudinal study could be accounted for either by gene patterns whose activation occurs regularly at a given chronological age or that were activated by a combination of environmental factors unique to the particular epoch monitored in the study. Similarly, environmental sources accounting for the phenotype could, in a given culture, occur regularly at a given age level, or in the particular instance be unique to the time interval during which the sample was followed, and thus involve completely nondevelopmental parameters.

The time-lag method compares two or more samples of subjects at the same age level measured at two or more points in time. Differences in the behavior of such samples here confound cohort membership (time of birth) and time-of-measurement. That is, differences in the behavior of samples of the same age measured at different historical points in time may be a function of environmental impact occurring during the interval separating the two time points, or due to generation membership; that is, differences in genetic activation patterns or early experience during the period when the older cohort had entered the environment, but the younger cohort did not yet exist. The time-lag method may be of particular interest if one is concerned with the stability across time of behaviors thought to be characteristic for a given chronological age level.

Since neither cross-sectional nor longitudinal methods can yield unambiguous measures of age differences (cross-sectional) or age changes (longitudinal), it has been suggested that new mixed strategies involving the replication of either cross-sectional or longitudinal studies be considered; such designs would then make it possible to segregate the variance due to chronological age from that attributable to cohort differences in cross-sectional studies and from nonage-related environmental effects in longitudinal studies.

Three different sequential strategies have been suggested: First, the cohort-sequential method, which involves the sampling of two or more cohorts at two or more times of measurement, permits segregation of variance due to age and cohort, under the assumption of trivial time-of-measurement effects. The cohort-sequential model can be seen as the replication of the traditional longitudinal study over a succession of cohorts. However, it does not absolutely require the collection of repeated measurement data, since the analyses of interest are possible also with data obtained from independent samples drawn at each scheduled time-of-measurement from the cohorts under investigation. As will be seen later on this model may be most useful for behavior genetic studies. A second strategy, the time-sequential method, requires sampling of two or more age levels at two or more times of measurement and permits segregation of age and time-of-measurement effects, under the assumption of trivial cohort differences. The model involves the replication of traditional cross-sectional studies. It is of particular interest to the educational and social psychologist as well as the developmentalist who wishes to segregate the effects of secular trends from developmental phenomena. The third strategy, the cross-sequential method, permits segregation of time-of-measurement and cohort effects, under the assumption of trivial age differences. This model can be applied to both repeated measurements on the same samples as well as to samples independently sampled at each time-of-measurement. The model is particularly useful for the study of behaviors that are not subject to developmental change once they become established, but that may be subject to change over cohorts as well as being influenced by transient environmental inputs. (For an heuristic example using these strategies, see Schaie 1972b).

Researchers interested in age change generally wish to find lawful relationships that describe ontogenetic change across the life course of a species (or a portion thereof). But such relationships can either generalize across cohorts or yield functions that may vary for successive cohorts in a predictable fashion. The conventional longitudinal method, however, can at best provide insight only for the specific cohort under study. Otherwise the cross-sequential method needs to be employed where stability of the phenomenon across age levels is expected, or the cohort-sequential method, when age changes are likely to obtain.

Researchers concerned with age differences, on the other hand, wish to determine in what manner individuals of different ages differ in behavior at one point in time. If it does not matter whether such difference is due to age or generational differences, then the cross-sectional method is perfectly applicable. But if we wish to know whether such differences are due to maturational or generation-specific events, then application of either the cross-sequential or time-sequential method will be required.

Environmental and Genetic Components of Developmental Change

An attempt is now made to relate the considerations discussed for developmental research strategies to the issue of proper attribution of environmental and genetic components as they might be expressed in heritability ratios or Cattellian type MAVA analyses. Indeed, in my original paper discussing a general model for the examination of developmental phenomena (Schaie 1965), it was intended to describe strategies that would differentiate maturational or ontogenetic change within individuals from differences between generations. Some confusion may have been introduced in that paper by the manner in which the term "maturational change" was used (see Baltes 1968; Buss 1973). I shall take the opportunity here to clarify this matter, since it may be directly relevant to the selection of appropriate data collection strategies in developmental behavior genetics studies.

The term "maturational change" as used by most developmentalists includes those genetic and environmental components that are age specific within the maturational schema of the species. Similarly the term "generational difference" must include both the difference in genetic patterns, if any, between successive generations as well as the differential impact of environmental experience upon the cohorts under comparison. But there is a significant difference. An age change (as conventionally measured by longitudinal comparison) involves environmental and genetic components that are age-related, the separation of which is often unsuccessfully attempted in twin studies. Simpler models, however, may be available (and will be discussed later on in this chapter), that would increase the power of such studies by permitting separation of the sum of age-related components from those environmental effects that are *not* age specific.

Cohort differences, on the other hand, assuming that two or more cohorts are compared over the same age range, may involve a genetic component arising out of some change in the gene pool or of gene activation and an environmental component, which in turn can be directly attributed to the sociocultural change occurring during the time interval when the more recent cohort had not yet come under scrutiny and that period when the oldest cohort had left scrutiny. Such input would normally not be age specific.

Given the validity of certain assumptions about the age specificity of developmental phenomena, there are circumstances where it is possible to identify time-of-measurement effects of strictly environmental origin and cohort effects of strictly genetic origin. Nevertheless, age or maturational effects will always confound environmental and genetic components, although their sum can be separated from general sociocultural impact (time-of-measurement effects).

Let us now examine the relation of environmental and genetic components within developmental change, particularly as they apply to differences between variance components and/or associated heritability ratios. We examine separately the simplest case where only main effects obtain as well as the likely more common case involving simple interactions. To begin with we need some definitions (see also Schaie 1970):

As indicated above we can define a developmental change (Lod) as determined from the longitudinal observation of the same organism as a composite of an age (Ad) and a time-of-measurement (Td) effect, such that

$$Lod = Ad + Td \tag{9.1}$$

Further, we can conceptually separate a difference (Csd) between groups at different developmental levels as determined from a cross-sectional study at one point in time into components associated with age and cohort (Cd) membership, such that

$$Csd = Ad - Cd \tag{9.2}{}^a$$

Also, we can consider a difference (Tld) between two population samples at identical developmental levels but observed at different points in time to consist of a cohort effect and a time of measurement effect such that

$$Tld = Td + Cd \tag{9.3}$$

Models Involving No Interaction Effects

Now let a pure age related effect Ad (as defined above) be partitioned into genetic and environmental components both accounting for observed changes from age m to age n:

$$Ad_{mn} = E_{mn} + G_{mn} \tag{9.4}$$

Let a cohort effect Cd be partitioned into a genetic component differentiating cohort i from cohort j and an environmental component. The latter component refers to that period of time when the older cohort is under scrutiny but the younger is not (because it either has not yet entered the environment or reached the age range of interest) extending from time $k - 1$ to k:

$$Cd_{ij} = G_{ij} + E_{ij} \tag{9.5}$$

Since time-of-measurement effects have been defined to be nonage specific, they can be expressed as a single environmentally determined component extending from time k to time ℓ, such that,

$$Td_{k\ell} = E_{k\ell} \tag{9.6}$$

We can now examine the components contributed by any of the three developmental research strategies as follows:

---

[a] The negative is the sign given to the cohort difference in equation (9.2) because the "oldest" cohort will have entered the environment first. For example, in the case of two age groups, equation (9.2) could be expanded as follows:

$$Csd_{1-2} = (A_1 + C_2) - (A_2 + C_1) = (A_1 - A_2) + (C_2 - C_1) = Ad_{1-2} - Cd_{2-1}$$

Longitudinal age changes for cohort i from age m to age n over the period from time k to time ℓ can be partitioned by inserting equations (9.4) and (9.6) into equation (9.1), and we obtain:

$$Lod_{i.mn} = E_{mn} + E_{k\ell} + G_{mn} \qquad (9.7)$$

Similarly, cross-sectional age differences between ages m and n at time k may be partitioned by inserting equations (9.4) and (9.5) into (9.2):

$$Csd_{mn.k} = G_{mn} - G_{ij} + E_{mn} - E_{ij} \qquad (9.8)$$

Time-lag changes occurring between time k and time ℓ at age m can also be partitioned by inserting equations (9.5) and (9.6) into (9.3):

$$Tld_{k\ell.m} = E_{k\ell} + G_{ij} + E_{ij} \qquad (9.9)$$

Thus far we have simply determined that longitudinal and time-lag differences can be separated conceptually into a genetic and two environmental components and that cross-sectional differences similarly can be thought of as consisting of two genetic and two environmental components. Let us next examine the possibilities of identifying the proportion of variance associated with the various environmental and genetic components, given the assumption that one of the three components of developmental change is trivial:

1. *Time-of-measurement differences trivial.* This assumption implies the absence of age-unrelated sociocultural impact. The assumption may be unreasonably strong in most human studies, but could obtain in animal studies or isolated societies where it might be possible to find de facto quasi-experimental isolation. Under this assumption it would follow that

$$Lod_{i.mn} = E_{mn} + G_{mn} \qquad (9.10)$$

that is, longitudinal changes would consist of age-specific environmental and genetic components. Next

$$Csd_{mn.k} = G_{mn} - G_{ij} + E_{mn} \qquad (9.11)$$

that is, cross-sectional differences involve genetic variance consisting of an age-specific component less a cohort-specific component and, of course, an age-specific environmental component. Further,

$$Tld_{kl.m} = G_{ij} \qquad (9.12)$$

indicating that time-lag differences, in this instance, will be limited to the genetic component of the respective cohort difference. Under these circumstances, cohort differences, whether identified by the time-lag method for a single-age level, or by the cohort-sequential method over many age levels, could reasonably be attributed to genetic differences between the cohorts under comparison.

Some readers will argue that changes in gene frequency must occur on an evolutionary time scale, and that differences in gene frequencies between narrowly defined population cohorts might therefore be largely attributable to sampling bias. But the effects of migration and selection pressures upon cohort differences within any given sampling frame may be much more substantial than most behavior geneticists and developmentalists have heretofore expected. For example, we find dramatic changes over a few decades in the ethnic composition of metropolitan populations and certain areas that have experienced massive migration due to technological revolutions such as the Appalachian mountains. In addition, the genetic component of the phenotypic differences between cohorts may be attributable to differential gene activation, caused by environmental factors, which might obviously occur over much shorter time lines than would be required for changes in gene frequency.

2. *Cohort differences trivial.* This model assumes that differences in the performance of two or more successive cohorts at the same age are not a function of cohort-specific experiences or gene activation patterns, but are due rather to the nonage-specific events that occurred over the particular time-of-measurement during which the observed age-related behavior change is thought to have occurred. Under this assumption it follows that

$$Lod_{i.mn} = E_{mn} + E_{kl} + G_{mn} \qquad (9.13)$$

The longitudinal changes here will consist of both the age-specific genetic and environmental components and a time-specific environmental component. Also,

$$Csd_{mn.k} = E_{mn} + G_{mn} \qquad (9.14)$$

cross-sectional differences will be limited to the age-specific genetic and environmental components. Further,

$$Tld_{kl.m} = E_{kl} \qquad (9.15)$$

since under the assumption of no cohort differences, the time-lag measure will be a direct estimate of the generalized time-specific environmental input. In this case we could then reasonably consider that time-of-measurement differences, whether identified by the time-lag approach for one age level, or by the time-sequential method across age levels, to indicate environmental differences, attributable to the specific input occurring over the monitored time interval. If the assumptions for the above model are met, a strategy is then available that would lend itself well to a different look at such problems, for example, as the effect of compensatory education on development. Here we would be able to differentiate between the component of variance attributable to the compensatory education (specific environmental input) and the maturational change (whether genetic and/or environmentally activated) occurring over a given age range.

3. *Age differences trivial.* The model assumes that differences in the performance of two or more successive cohorts at two or more measurement points are not a function of maturational events but rather attributable to differences in gene activation and or experience between the two cohorts as well as differential environmental exposure over time. The model may be particularly appropriate for the study of behavioral development past adolescence. Given the above assumptions it follows that

$$Lod_{i.mn} = E_{kl} \qquad (9.16)$$

longitudinal changes will be the specific function of generalized environmental input occurring to organisms of all ages over the time span monitored. Next,

$$Csd_{mn.k} = G_{ij} + E_{ij} \qquad (9.17)$$

that is, the cross-sectional difference will reflect both genetic and environmental components that differentiate the cohorts. And

$$Tld_{kl.m} = G_{ij} + E_{kl} + E_{ij} \qquad (9.18)$$

the time-lag difference reflects the cohort difference plus the environmental nonage-specific effect occurring over the time monitored. In this instance also, time-of-measurement differences attributable to environmental variance could be ascertained, whether measured by traditional longitudinal study or via the cross-sequential method.

This model is, of course, of limited interest to developmentalists, but it may be useful nevertheless in the description of nondevelopmental population traits, and the identification of variance components that are clearly attributable to environmental factors subject to distinctive secular trends.

To summarize, given the absence of significant interactions, it is possible to segregate a component G_{ij} denoting the proportion of variance due to genetic differences between cohorts under the assumption that sociocultural change does not operate upon the variable of interest, and a component E_{kl} denoting the proportion of variance due to the generalized effect of sociocultural change over a specific epoch, either under the assumption of no age-specific changes or the assumption of no cohort differences.

Models Allowing Interaction Between Two Components of Developmental Change

Let us look next at what happens when the restriction upon interactions is relaxed. The interaction model, of course, cannot be studied for single cohort longitudinal, simple cross-sectional or single-age time-lag designs, but only for sequences thereof.

1. *Time-of-measurement differences trivial.* The most interesting case

here is the model where time-of-measurement variation is assumed to be trivial, since this model would seem to permit simultaneous identification of certain unconfounded environmental and genetic components. More specifically we would find that, under the assumptions given,

$$Ad_{mn.ij} = E_{mn} + G_{mn} \qquad (9.19)$$

$$Cd_{ij.mn} = G_{ij}, \quad \text{and} \qquad (9.20)$$

$$Ad_{mn.ij}Cd_{ij.mn} = E_{mn.ij} + G_{mn.ij} - G_{ij.mn} = E_{mn.ij} \qquad (9.21)$$

Omega squared obtained from the cohort-sequential analysis of variance would therefore provide us with an estimate of the component due to age-specific variation regardless of cohort membership (Ad), the component of variation due to genetic differences between cohorts, specific to the observed cohorts but regardless of age level (Cd), and finally the component of variation due to age-related environmental impact specific to the cohorts under examination ($Ad \times Cd$ interaction).

The cohort-sequential model would seem to be a rather powerful approach to behavior genetic analysis for those traits that whether through genetic activation or early behavioral imprinting are resistant to additional transformations occurring via secular trends. In this model we can examine jointly intraindividual changes (Ad), interindividual differences (Cd) and their interaction. Moreover, we can specify that intraindividual change must confound genetic and environmental factors, while interindividual differences (in cohort membership) can be logically attributed to differences in genetic activation patterns and the interaction of intra and interindividual differences can be attributed to environmental events.

2. *Cohort Differences trivial.* Under this assumption we would find that

$$Ad_{mn.kl} = E_{mn} + G_{mn} \qquad (9.22)$$

$$Td_{kl.mn} = E_{kl}, \quad \text{and} \qquad (9.23)$$

$$Ad_{mn.kl}Td_{kl.mn} = E_{mn.kl} + G_{mn.kl} + E_{kl.mn} \qquad (9.24)$$

But, age-specific genetic variation should not be affected by transitory environmental events and $G_{mn.kl}$ should therefore be zero. Simplifying and collecting terms, we can therefore write:

$$Ad_{mn.kl}Td_{kl.mn} = 2E_{mn.kl} \qquad (9.25)$$

Again, omega squared from the appropriate time-sequential analysis of variance would provide an estimate of variance due to age-specific change regardless of the period over which it occurs (Ad), variation due to the environmental component occurring during the specific period regardless of age level at which observed (Td), and an interaction term indicating the component due to age-specific environmental variation occurring as a function of the specific periods during which observations are made.

As already indicated above, this model may be particularly useful for the study of policy issues such as compensatory education. The interaction model is more powerful than the simple model, because in addition to estimating the component of variance due to time-specific environmental input, the interaction term will also provide information on the proportion of such time-specific input that has differential effect depending upon the age of subjects at the time of intervention.

3. *Age differences trivial.* Under this assumption, finally, we have

$$Cd_{ij.kl} = G_{ij} + E_{ij} \tag{9.26}$$

and

$$Td_{kl.ij} = E_{kl} \tag{9.27}$$

Further,

$$Cd_{ij.kl}Td_{kl.ij} = G_{ij.kl} + E_{ij.kl} + E_{kl.ij} \tag{9.28}$$

But since cohort-specific genetic variation should not be differentially effected by transitory environmental events $G_{ij.kl}$ should be zero. Simplifying and collecting terms, we therefore have

$$Cd_{ij.kl}Td_{kl.ij} = 2E_{kl.ij} \tag{9.29}$$

In this case the appropriate omega squared from the cross-sequential ANOVA will yield a component representing cohort specific genetic variance plus time specific environmental variance (representing the period when the younger cohort had not yet entered the environment) (*Cd*), another component that estimates the nonage-related but time-specific environmental variation (*Td*), and the interaction component that represents time-specific environmental effect specific to the cohorts under examination (*Cd* × *Td* interaction).

This model may be of interest for assortative mating, parent-offspring, and other population studies, since it permits estimation of the magnitude of behavior differences that can be attributed to time-specific environmental factors as well as those time-specific factors that express themselves differentially for successive cohorts.

Estimation of Developmental Heritability Ratios

Traditional methods of quantitative genetics related to the estimation of heritability ratios seem to have concerned themselves entirely with the problem of whether or not a given trait at one point in time may be found for samples bearing differential genetic relationships to one another. Such all or none approach contrasts with the interest of the developmentalist who wishes to study both emergence and change in behavioral characteristics. Thus, the entire issue raised by A. R. Jensen's (1969) controversial contribution may be developmentally irrelevant. While it may be of theoretical interest to know what the relative contribution of genetic and environmental variance might be at a given age for a particular point in time (and this is all the twin studies really tell us) it is of much more concern to what extent developmental *change* can be accounted for as a function of environmental and preprogrammed maturational factors. Policy consequences drawn from behavior genetic studies, however, would have a much sounder base if they would rely upon data relevant to the issue of developmental change. Jensen (1973) has attempted to address himself to these issues but the methodology used in his analysis is open to challenge for reasons indicated in this chapter.

Studies of assortative mating and of parent-child or sibling relationships make the strong assumption that neither age nor cohort effects are operating upon the variables of interest in cross-sectional studies. In longitudinal studies of this type it is similarly assumed that age and time-of-measurement effects can be safely ignored. As indicated in our discussion it may be possible to segregate developmental variance with far less restrictive assumptions, and I shall now recommend a series of ratios that might be useful to behavior geneticists interested in addressing the question of the genetic basis of developmental change. The first group of three ratios addresses itself to identifying the relative proportion of variance due to age, time, and cohort effects. These ratios might be usefully applied to twin as well as other data. Next, a new type of heritability ratio is proposed, which could be computed from data matrices available in many developmental studies now in the literature. Also, data collections planned for such analyses would be far less expensive than those for other currently advocated techniques, and might address themselves directly to issues such as those posed by Cattell (1970).

1. *Age-specific versus cohort-specific variance.* From the cohort-sequential ANOVA we can estimate an age/cohort ratio such that

$$AC^2_{mn.ij} = \frac{\omega^2 A d_{mn.ij}}{\omega^2 A d_{mn.ij} + \omega^2 C d_{ij.mn}} \qquad (9.30)$$

Under the assumption of trivial time-of-measurement variance, this ratio will give the relative proportion of nonerror variance attributable to age-specific sources from ages m to n for cohorts i and j. The ratio contrasts the effect of intra and interindividual change over time. This ratio might also be useful in comparing the relation between parents and offsprings when data are available on behavior change over the same age range for both generations.

2. *Age-specific versus time-specific variance.* Similar to the age/cohort ratio, we can derive an age/time ratio from the time-sequential ANOVA, such that

$$AT^2_{mn.kl} = \frac{\omega^2 Ad_{mn.kl}}{\omega^2 Ad_{mn.kl} + \omega^2 Td_{kl.mn}} \quad (9.31)$$

Assuming trivial cohort variance, this ratio would estimate the relative proportion of nonerror variance due to age-specific sources for ages m and n across times k and l. This ratio would be useful in the study of sibling relationships, where behavior change over the comparable developmental period would occur at different points in time.

3. *Time-of-measurement versus cohort-specific variance.* From the cross-sequential ANOVA, we can next estimate a time/cohort ratio, such that

$$TC^2_{kl.ij} = \frac{\omega^2 Td_{kl.ij}}{\omega^2 Td_{kl.ij} + \omega^2 Cd_{ij.kl}} \quad (9.32)$$

Assuming age variance to be trivial, this ratio estimates the relative proportion of nonerror variance due to nonage-related time-specific effects at times k and l across cohorts i and j. This ratio is of interest when we are concerned with determining the proportion of variance due to the environmental impact in a given time period as against the joint genetic and environmental impact specific to cohort membership. This information may be useful in relation to issues such as compensatory education. It would permit the proportion of variance effected by compensatory efforts introduced over the time period monitored to be contrasted with the difference between successive cohorts attributable to the combination of differential genetic and early socialization patterns.

4. *Cohort-specific heritability ratio.* From the cohort-sequential ANOVA it is furthermore possible to determine what may well be a truly developmental ratio, a ratio that will contrast the environmental and genetic components differentiating two successive cohorts. From equations (9.20) and (9.21) we find that

$$h^2_{ij.mn} = \frac{\omega^2 Cd_{ij.mn}}{\omega^2 Cd_{ij.mn} + \omega^2 Ad_{mn}Cd_{ij}} \quad (9.33)$$

The ratio gives the relative proportion of genetic variance differentiating cohort i and j at ages m and n. This ratio may be more suitable in addressing the issues raised by Cattell (1970) than the alternate procedures suggested by Buss (1973).

References

Baltes, P.B. 1968. Longitudinal and cross-sectional sequences in the study of age and generation effects. *Human Development* 11: 145-71.

Buss, A.R. 1973. An extension of developmental models that separate ontogenetic changes and cohort differences. *Psychological Bulletin* 80: 466-79.

Campbell, D.T., and Stanley, J.C. 1963. Experimental and quasi-experimental designs for research. In *Handbook of research on teaching*, ed. N.L. Gage. Chicago: Rand McNally.

Cattell, R.B. 1970. Separating endogenous, exogenous, ecogenic and epogenic component curves in developmental data. *Developmental Psychology* 3: 151-62.

———. 1973. Unraveling maturational and learning development by the comparative MAVA and structured learning approaches. In *Life-span developmental psychology: Methodological issues*, eds. J.R. Nesselroade and H.W. Reese. New York: Academic Press.

Jensen, A.R. 1969. How much can we boost I.Q. and scholastic achievement? *Harvard Educational Review* 39: 1-123.

———. 1974. Cumulative deficit: A testable hypothesis? *Developmental Psychology* 10: 996-1019.

Nesselroade, J.R., and Baltes, P.B. 1974. Adolescent personality development and historical change: 1970-72. *Monographs of the Society for Research in Child Development* 39, no. 154.

Schaie, K.W. 1965. A general model for the study of developmental problems. *Psychological Bulletin* 64: 92-107.

———. 1970. A reinterpretation of age-related changes in cognitive structure and functioning. In *Life-span developmental psychology: Research and theory*, eds. L.R. Goulet and P.B. Baltes. New York: Academic Press.

———. 1972a. Can the longitudinal and method be applied to psychological studies of human development? In *Determinants of behavioral development*, eds. F.Z. Moenks, W.W. Hartup, and J. de Wit. New York: Academic Press.

———. 1972b. Limitations on the generalizability of growth curves of intelligence: A reanalysis of some data from the Harvard Growth Study. *Human Development* 15: 141-52.

———. 1973. Methodological problems in descriptive research on adulthood and aging. In *Life-span developmental psychology: Methodological issues*, eds. J.R. Nesselroade and H.W. Reese. New York: Academic Press.

———. 1974. Translations in Gerontology—from lab to life: Intellectual functioning. *American Psychologist* 29: 802-7.

———. 1976. Quasi-experimental designs in the psychology of aging. In *Handbook of the the Psychology of Aging,* eds. J.E. Birren and K.W. Schaie. New York: Reinhold-Van Nostrand. In press.

Schaie, K.W., and Gribbin, K. 1975. Adult development and aging. *Annual Review of Psychology* 26: 65-96.

Wohlwill, J.R. 1973. *The study of behavioral development.* New York: Academic Press.

10 Possible Directions for Developmental Human Behavior Genetics

Irving I. Gottesman
University of Minnesota

Congratulations are in order to the organizing committee for the workshop on developmental human behavioral genetics that led to this book. It takes courage for a National Institutes of Health (NIH) Study Section to expose itself to the burden of acquiring new knowledge and skills in an area basic to its grant evaluation ability, since it admits the possibility that study sections are not omniscient—a sentiment expressed by many a rejected applicant. The burgeoning of knowledge in the behavioral sciences threatens us all with rapid obsolescence, but the problem is exacerbated for those of us involved in the developmental aspects since the task of keeping up means keeping abreast of many domains usually kept separate by tradition and training. It is inevitable that an intense exposure to the foundations for thinking in a developmental-behavioral-genetics fashion will lead to information overload. Hopefully, the exposure will have also led to a whetting of appetites for pursuing these ideas at a more leisurely pace; for this purpose a sampler of some resources, necessarily idiosyncratic, directed at developmental psychologists desiring a more profound background is given in an appendix to this chapter. Whenever an assembly of experts is asked to do "their thing," the processes of condensation and simplification lead to a false view of the degree of clarity and consensus that exists. Even-handed as they were, our panel of expert witnesses and discussants could often only hint at the complexity underlying their summary statements; we lack the magical power to transform the reader into cognoscenti with the contributions prepared for two days' worth of workshops, but heightening your consciousness is our sufficient reward.

There is no point in trying to hide the fact that a mixture of legitimate and illegitimate contenders have contributed to a certain militant approach to the knowledge about human behavior genetics and its interpretation. I can only caution the consumer to beware of the radicalization and the politicization of data used in the service of ideology rather than in the service of science in its quest for understanding. Complexity and uncertainty are inherent in the domain of developmental human behavior genetics and cannot be evaded. Thus, there is no conspiracy to suppress forbidden knowledge about man's genetic architecture, but rather an obstinacy in those who refuse to recognize the technological barriers that currently limit the so-called genetic comparisons between populations for behavioral characteristics. The enormous gaps in knowledge about the events between gene action and neurophysiology and between neurophysiology and adaptive behavior are seldom made explicit when researchers describe efforts within their territories. Such gaps lead to the uncritical insertion of soft psychometric data into the elegant formulas of population genetics with the results being treated as if they constituted "biological truths." Recognition of such frailties and others expressed by the candid comments in the preceeding chapters is *not* tantamount to licensing regression to a naive Watsonian behaviorism or to Lysenkoist distortions about genetics or to obscurantism in general.

As many of the contributors to this book have made clear, there is hardly yet a developed, mature discipline of *developmental* human behavior genetics but it is certainly an emerging discipline; the concept of development is a difficult one both for ontogenetic psychology and for developmental biology. In a beautiful essay entitled "Form, End and Time" in his book *The Strategy of the Genes,* C. H. Waddington gave an overview of the conceptual framework with which we find ourselves grappling. An adequate picture of any human can only be provided by considering the effects on it of three different types of temporal change, each being effective simultaneously and continuously:

The three time-elements in the biological picture differ in scale. On the largest scale is evolution; any living being must be thought of as the product of a long line of ancestors and itself the potential ancestor of a line of descendants. On the medium scale, an animal . . . must be thought of as something which has a life history. It is not enough to see that horse pulling a cart past the window as the good working horse it is today; the picture must also include the minute fertilized egg, the embryo in its mother's womb, and the broken-down old nag it will eventually become. Finally, on the shortest time-scale, a living thing keeps itself going only by a rapid turnover of energy or chemical change; it takes in and digests food, it breathes, and so on.

In the biological picture towards which we are finding our way, the three time systems will have to be kept in mind together. That is the feat which common sense still finds difficult. Even in current biology, most of our theories are still only partly formed because they leave one or the other of the time scales out of account [Waddington 1957].

Therein lies one of the challenges to developmental psychologists who have largely been working within the medium time scale; that is, become aware of large-scale time (roughly equated with evolution and population genetics); remind yourselves of the importance of embryology and gerontology; and keep alert to the potential relevance of physiology and developmental genetics.

An informative example of some of the congruences between the ideas above and reality is provided by a close look at the behavior of milk drinking in mammals (Gottesman and Heston 1973). Mother's milk or some equivalent provides all the nutrients necessary for the growth of mammals during the early part of infancy. Rarely, a child will be allergic to the protein in milk or have the genetically determined disorder known as galactosemia; such children require a nonmilk food source from the beginning of life. It is now clear that many adults and older children (after ages two to eight or so) from all races but especially non-Caucasian populations have a considerably lower tolerance than others to the milk sugar lactose, because of low levels of intestinal lactase activity. (Lactase is the enzyme responsible for the splitting of lactose into soluble sugars that provide, then, 30 to 60 percent of the calories present in milk. Lactose is not absorbed by the small intestine.) These same individuals with "later onset" lactose intolerance had no difficulty at all using milk as infants.

In all mammals so far studied except man lactase activity is high in infancy and drops to near zero after weaning. This state of affairs can probably be taken as the normal mammalian pattern; in prevalence studies 60 to 100 percent of many non-Caucasian populations showed this pattern. In contrast, however, only 0 to 20 percent of many North European populations and their descendants showed low levels of lactase activity. The question of whether the populations differ in this respect for genetic reasons after 10,000 years of natural selection or rather as the result of continuing to drink milk after weaning is probably answered by the evolutionary argument.

A cross-sectional picture of the levels of lactase available to the rat as a function of prenatal and postnatal age is informative. The "wisdom of the body" as it was metaphorically called by the famed W. B. Cannon a half-century ago is revealed by the evolutionarily optimized regulation of lactase levels so as to be peaking just when the rat baby will start to suckle. We can reason that since it would be inefficient for the gene to be switched on too early or to stay on after its product was no longer needed, that was the way the system evolved. A similar developmental pattern for this enzyme is observed in primates other than man and in the majority of humans with certain informative exceptions.

Since the consequence of low lactase levels is the inability to digest milk followed by such symptoms as diarrhea and gastrointestinal pain, we would expect lactose intolerance to influence patterns of milk consumption among children as the operon turns off. The changes in age (cross-sectional data) in the prevalence of a normal response to a lactose tolerance test in different populations are given in Gottesman and Heston (1973). The data lead to the conclusion that it is not the scarcity of milk in the postweaning diet that leads to low levels of lactase, although its availability may divert the ball on the epigenetic landscape and delay milk intolerance temporarily. The populations that consume milk as adults are, by and large, of North European origin with exceptions that test the rule; the Masai and pastoral Fulani of Africa, both of which groups have raised cattle and consumed milk for thousands of years, are very good milk drinkers without lactase deficiency.

We can speculate that selection for lactose tolerance must have begun ten to twelve thousand years ago when human populations began domesticating milk-producing animals. Because the adult form of intolerance is not fatal and would only be disadvantageous when food supplies were very marginal, selection pressures must have been gentle. We may also note that selection favoring tolerance must have increased in populations where significant numbers had already become tolerant: The possession of a favorable trait increasing fitness leads to displacement of other phenotypes. Once some members of a population utilized milk as a food, the remaining members were at a somewhat increased disadvantage.

As in other examples of interaction between environment and genes the more one understands about this specific phenomenon, the more difficult it becomes to separate genes from environment. In the case of lactase it appears that a cultural-technological advance, domestication of animals, was intermeshed with a change in gene frequency.

As suggested by data from monkeys, primitive man, like all mammals, must have been lactose intolerant after infancy. It is tolerance of lactose that must have evolved. What magnitude of selective advantage would have been required to change the frequency of a favorable dominant mutation to currently observed levels? Accepting the current prevalence of lactase deficiency in contemporary intolerant populations to be 90 percent as opposed to a 10 percent in northwestern Europe, the corresponding frequencies of the gene for adult lactase production would be .05 in intolerant populations and about .60 in tolerant populations. We have worked out the approximate selection intensity against homozygote nonlactase producers required to change the gene frequency from .05 to .60 in the 400 generations since domestication of sheep and goats. The selection intensity is approximately .01. The literal meaning of this number is that if lactose tolerant persons had an average of 1 percent more children surviving per generation than lactose intolerant persons, the observed change in phenotype frequency could occur in the time available. The value .01 is commonly encountered and is of a reasonable magnitude.

Concluding Remarks

Both the backgrounds of the contributors to this book and the subject matters they have broached show how diverse is the informational mosaic that goes into the make up of what we are now calling human developmental behavior genetics. Heuristic integration is an enormous task and will require multidisciplinary study sections evaluating multidisciplinary research applications. The necessity and opportunities for developmental genetics and developmental psychology to both contribute to the solutions to society's problems is clarified by the contents of this book. The organizers and the contributors can only hope that their efforts will reach a wide and influential audience.

References

Gottesman, I.I., and Heston, L.L. 1973. Summary of the Conference on Lactose and Milk Intolerance. Washington, D.C.: DHEW Publ. No (OCD): 73-19.

Waddington, C.H. 1957. *The Strategy of the genes*. London: Allen & Unwin.

Appendix 10A
A Sampler of Resources in Behavioral Genetics for Developmental Psychologists

Journals

Annual Review of Psychology

American Journal of Human Genetics

Behavior Genetics

Human Heredity

Social Biology

Books

Cavalli-Sforza, L.L., and Bodmer, W. 1971. *The genetics of human populations*. New York: Freeman.

Dobzhansky, T. 1962. *Mankind evolving*. New Haven: Yale University Press.

Ehrman, L.; Omenn, G.S.; and Caspari, E., eds. 1972. *Genetics, environment, and behavior*. New York: Academic Press.

Freedman, D.G. 1974. *Human infancy: An evolutionary perspective*. New York: Wiley.

Fuller, J.L., and Thompson, W.R. 1960. *Behavior genetics*. New York: Wiley.

Glass, D.C., ed. 1968. *Genetics (Biology and Behavior Series)*. New York: Rockefeller.

Gottesman, I.I. and Shields, J. 1972. *Schizophrenia and genetics: A twin study vantage point*. New York: Academic Press.

Hirsch, J., ed. 1967. *Behavior-genetic analysis*. New York: McGraw-Hill.

McClearn, G.E., and DeFries, J.C. 1973. *Introduction to behavioral genetics*. New York: Freeman.

Rosenthal, D. 1970. *Genetic theory and abnormal behavior*. New York: McGraw-Hill.

Scott, J.P., and Fuller, J.L. 1965. *Genetics and the social behavior of the dog.* Chicago: University of Chicago Press.

Slater, E., and Cowie, V. 1971. *The genetics of mental disorders.* Oxford: Oxford University Press.

Vandenberg, S.G., ed. 1968. *Progress in human behavior genetics.* Baltimore: Johns Hopkins University Press.

Glossary

acetylation: The introduction of an acetyl group into the molecule of an organic compound.

adrenogenital syndrome: The primary defect is a genetic one, transmitted as a recessive, which prevents the synthesis of cortisol in the adrenal cortex. It causes a defect of the genital anatomy if the fetus is female.

albinism: A condition characterized by a lack of pigment in skin, hair, eyes, etc. It is a recessive trait occurring in about 1 in 10,000 persons. The genetic block involves a step between the amino acid tyrosine and the pigment melanin.

allele: One of two or more alternative forms of a gene occupying the same locus on a particular chromosome.

amino acids: The building blocks of proteins.

amniocentesis: A procedure by which fluid is taken from the amniotic sac surrounding the fetus permitting prenatal diagnosis of chromosomal errors and of some inborn errors of metabolism.

antibody: An immunoglobulin molecule that interacts only with the antigen that induced its synthesis in lymphoid tissue.

assortative mating: Nonrandom mating. A tendency for members of a population to mate with others of a similar (positive assortative mating) or dissimilar (negative assortative mating) phenotype.

autosome: Any chromosome except the sex chromosomes.

backcross: Mating between a hybrid and one of the two parental types, for example, Aa × aa.

Barr body: The sex chromatin spot in the nucleus of cells containing more than one X chromosome.

bilirubin: A bile pigment.

biochemical genetics: A branch of genetics dealing with the chemical nature of hereditary determinants and the manner of their action in development and function.

carrier: An individual heterozygous for a recessive allele.

choreoathetosis: A condition marked by the occurence of a wide variety of involuntary, rapid, highly complex jerky movements.

chromotography: A means of separating and identifying the components of mixtures of molecules.

concanavalin A: A compound that will agglutinate mammalian blood.

concordance: Presence of a particular condition in both members of a twin pair.

congeneic: A strain of animals differing genetically from another only at one locus, or one portion of a chromosome.

co-twin control method: The effects of application of a specific environmental treatment to one member of a monozygotic twin pair.

cretinism: A chronic condition due to a congenital lack of thyroid secretion, marked by arrested physical and mental development and lowered basal metabolism.

cytogenetics: A field of investigation concerned with the correlation of genetic and cytological systems.

dihybrid: An individual which is heterozygous with respect to two pairs of alleles.

dimer: A molecule formed by combination of two simpler subunits.

DNA: Deoxyribonucleic acid.

Down's Syndrome: A type of mental retardation due to chromosome abnormality. All patients have all or most of chromosome 21 represented three times instead of twice. The clinical features, in addition to mental retardation include a peculiarity in the eyelid folds, short stature, stubby hands and feet, and congenital malformations, especially of the heart.

dominant: An allele that produces a phenotypic effect when heterozygous or homozygous.

ectopic: Aberrant, out of place.

electrophoresis: The movement of charged particles under the influence of an applied electric field.

eosinophilia: An excess in the number of eosinophil cells which are medium-sized leukocytes (white blood cells).

erythrocytes: Red blood cells.

eugenics: The study of methods of improving genetic quality of the human species.

euphenics: The amelioration of genetically influenced defect by appropriate environmental treatment.

euthenics: The study of the establishment of optimal environmental conditions.

F_1, F_2: First filial generation; offspring resulting from a given mating. The second filial generation resulting from crossing of F_1 individuals.

full sibs: Individuals that have both parents in common.

galactosemia: An autosomal recessive disorder of galactose metabolism characterized by vomiting, diarrhea, jaundice, poor weight gain, malnutrition and mental retardation: symptoms regress if galactose is removed from the diet.

gene: A unit of genetic material localized in the chromosome. The gene is operationally defined as a specific sequence of nucleotides (in DNA or RNA) acting as a functional unit.

gene frequency: The proportion of one particular allele to the total of all alleles at this genetic locus in a breeding population.

gene pool: The total genetic information coded in the total of all genes in a breeding population at a given time.

genetic drift: Changes in gene frequencies due to random fluctuations.

genome: Chromosomal set of a gamete consisting of the sum total of genes.

genotype: Genetic constitution of an individual.

glycolysis: The energy-yielding breaking down of glucose to lactic acid.

glycoproteins: A class of compounds of protein with a carbohydrate.

glyconeogenesis: The formation of carbohydrates from molecules that are not themselves carbohydrates, such as amino acids or fatty acids.

half-sibs: Individuals that have only one parent in common.

hemizygous: Situation in which only one allele of a locus is present, as in sex-linkage in the heterogametic sex or in the case of loss of chromosome segments.

heritability: The ratio of genetic variance to total phenotypic variance (narrow definition).

heterozygous: The condition of having different alleles at a genetic locus.

histidinemia: An autosomal recessive condition characterized by elevated blood histidine levels, sometimes mental retardation, speech defect.

homocystinuria: An inborn error of metabolism characterized by excretion of homocystine in urine.

homologous: Chromosomes having the same sequence of genes.

homozygous: The condition of possessing identical alleles at a given locus, e.g., AA or aa.

Huntington's Chorea: A dominant condition with a variable age of onset, characterized by loss of motor control, progressive dementia, and mental impairment.

hybrid: Any offspring of a cross between two genetically unlike individuals.

hybridization: A mating of two genetically different individuals.

hydroxylation: The chemical addition of the OH^- radical.

hyperuricemia: Excess of uric acid in the blood.

inborn error of metabolism: A genetically determined biochemical disorder in which a specific enzyme defect produces a metabolic block.

inbred line: A more or less homozygous line derived from an out-breeding population by repeated inbreeding. In mice, a line is regarded as inbred only after 20 consecutive generations of sib-mating.

inbreeding: The mating of relatives.

isoallele: An allele so similar in its phenotypic effect to that of other alleles as to require special techniques for identification.

isozyme: Any of a number of multiple molecular forms of an enzyme found in differentiated cells of the same individual.

Karyotype: Somatic chromosomal complement of an organism.

Kleinfelter's Syndrome: A syndrome including small testes, over-development of the mammary glands and elevated urinary gonadotrophins. Individuals with XXY karyotype usually have Kleinfelter's Syndrome.

Lesch-Nyhan Syndrome: An X-linked deficiency of the enzyme hypoxanthine-guanine phosphoribosyl-transferase with the syndrome of choreopathetosis, self-mutilation of fingers and lips, hyperuricemia and mental retardation.

linkage: The non-random assortment of genes due to their location on the same chromosome.

locus: The fixed position of a gene on the chromosome.

MAVA: "Multiple abstract variance analysis" developed by Raymond Cattell, leading to the estimation of nature:nurture ratios, and an assessment of the importance of genetic and environmental influences within the family as well as within the culture.

maternal inheritance: Inheritance controlled by extrachromosomal (cytoplasmic) hereditary factors.

Mendelian inheritance: The mode of inheritance of chromosomal genes in contrast to cytoplasmic hereditary determinants.

messenger RNA: RNA that is synthesized in the nucleus on a DNA template, then passes to the cytoplasm, attaches to one or more ribosomes, and serves as a template on which amino acids are assembled to form a polypeptide chain.

metachromatic leukodystrophy: A disorder transmitted as an autosomal recessive trait characterized by an accumulation of sulfatide in neural and non-neural tissues, with a diffuse loss of myelin in the central nervous system.

methylated: Containing or combined with a methyl group.

miscegenation: The intermarriage or interbreeding of persons of different races.

monogenic: Referring to phenotypic differences controlled by the alleles of one gene.

mosaicism: The presence within a single individual of cells differing with respect to chromosome structure, number, or genotype.

mutant: Resulting from mutation.

mutation: Heritable change in the genetic material.

neuroblastoma: Neoplasm (tumor) of nervous system origin composed chiefly of neuroblasts.

nucleotide: The combination of a purine or pyrimidine base, a sugar, and a phosphate group.

operon: A group of closely linked structural genes with related function that are turned on and off in concert under the control of a controlling gene, or operator.

peptide: A compound with two or more amino acids. Peptides join to form polypeptides, which in turn join to form protein.

pedigree: A record of ancestry or genealogical register, frequently presented as a diagram for two or more generations.

phenotype: An observable structural or functional property of an individual.

Phenylketonuria: A genetic defect in metabolism resulting in mental retardation. The metabolic block occurs in the conversion of phenylalanine to tyrosine. The metabolic derangement is attributed to the inactivity of the enzyme phenylalanine hydroxylase.

picogram: A metric unit of mass (weight), being 10^{-12} gram.

pleiotropy: The influence by one gene on two or more phenotypes.

polygenic: Term used to describe multiple genetic factors.

polygenic inheritance: Inheritance in which F_1 offspring are intermediate between their parents with respect to some trait, and F_2 offspring vary widely.

polymorphism: Literally, having many forms; the regular occurrence in the same population of two or more genotypes.

proband: In human genetics, the "affected" person with whom the study of a particular character in a family begins.

PTC: Phenylthiocarbamide—A chemical that is tasteless to individuals homozygous for a particular recessive allele but is bitter to individuals who are heterozygous or homozygous for the dominant allele.

races: Populations that differ in the frequencies of certain genes in their gene pool.

recessive: An allele that produces a phenotypic effect only when homozygous.

recombination: Combinations of alleles in offspring different from parental combinations. Recombination occurs as a consequence of exchange of chromosomal material between homologous chromosomes.

renaturation: In the case of DNA, return to the double helix configuration.

RNA: Ribonucleic acid.

ribosomal RNA: Nucleic acid component of ribosomes.

ribosome: Cellular components made up of ribosomal RNA and several kinds of proteins that are the primary sites of amino acid polymerization during protein synthesis.

sex chromosomes: Chromosomes that are non-homologous in one sex (X and Y chromosomes in man).

sex-linked: Determined by a gene located on the X chromosomes. (Y linkage in man is extremely rare).

sickle-cell anemia: A severe disease in which the red blood cells become crescent-shaped, that is associated with impaired oxygen transport.

sulfatide: One of a class of cerebroside sulfuric esters found largely in the medullated nerve fibers, that may accumulate in the white matter of the brain in metachromatic leukodystrophy.

Tay-Sachs Disease: An autosomal recessive condition where lipid metabolism is disrupted. Also known as infantile amaurotic idiocy. Homozygous individuals show an absence of a component of the enzyme beta-D-N-acetylhexosaminidase. Although normal in appearance at birth, nystagmus and paralysis begin within a few months and paralysis continues until death (usually before the age of two).

Testicular Feminization Syndrome: (46, XY karyotype) Represents a well-defined form of male pseudohermaphroditism. Persons are legally female with well-developed breasts, sparse pubic hair, and external genitals are female but with a blind vagina. Testes of almost normal size are found in the abdominal cavity or the inguinal canals.

transfer RNA: A form of RNA found in the cytoplasm that transports amino acids to ribosomes where proteins are assembled.

Turner's Syndrome: Symptom patterns include short stature, a variety of somatic malformations, sexual infantilism, "streak gonads," and a space-form deficit. Persons with XO karyotype suffer from the condition.

X-inactivation: The hypothesized inactivation in somatic cells of all X chromosomes in excess of one.

Indexes

Author Index*

Aebi, H., 100, *109*
Allen, G., 9, *21*
Amano, T., 108, *108*
Anderson, G., 37n
Anderson, V.E., **113-118**, 45, *53,* 105, 107, *108*, 117-118
Appel, S.H., 37, *39*
Atkinson, G.F., 44, *53*
Ayd, F.J., Jr., 185, *194*

Bailey, D.W., 33, 37, *38, 39*
Bakay, B., 105, *110*
Ball, E.D., 99, *108*
Baltes, M.M., 11, *21*
Baltes, P.B., 11, *21,* 34, 35, *38,* 205, 209, *218*
Barchas, J., 131, *137*
Barondes, S.H., 99, 100, *109, 111*
Beecher, H.K., 183, 184, 190, *194*
Beerstecher, E., Jr., 141, *144*
Belmont, L., 19, *21,* 55, 57, 62, 127, *136*
Beloff, J.R., 51, *52*
Bergsma, D., 189, *194*
Berkeley, G., 189, *195*
Berman, J.L., 121, *122*
Berry, H.K., 141, *144*
Berry, J.S., 141, *144*
Bilbro, W.C., 44, *54*
Blewett, D.B., 51, *52*
Blomstrand, C., 100, *110*
Blum, J.E., 15, *23,* 35, *38,* 86, 88, 89, *90*
Boadle-Biber, M.C., 113, *118*
Bock, R.D., 50, *52,* 87, *90*
Bodmer, W., 125, 132, *136,* 227
Borgaonkar, D.S., 187, 189, *194*
Boyce, A.J., 69, *74, 75*
Brady, R.O., 100, *110*
Breland , H.M., 55, 58, *62*
Brenner, S., 115, *117*
Broman, S.H., 45, 51, *53*
Bruhl, H.H., 116, *118*
Bruner, J., 20
Bulmer, M.G., 47, *52*
Burt, C., 10, 16, *21,* 22, 47, *52*
Buss, A.R., 205, 209, 217, *218*

Cain, D.F., 99, *108*
Callahan, D., 189, *195*
Campbell, D.T., 205, *218*
Campbell, E.Q., 61, *62*
Campbell, M.A., 129, *137*

Cannon, W.B., 223
Caplan, R., 99, *109*
Capron, A.M., 183, 184, *194*
Carrivick, P.J., 69, *74, 75*
Caspari, E., 25, *30, 227*
Castelano, C., 33, *38*
Cattell, R.B., 10, 11, 17-19, *22,23,* 50, 51, *52,* 205, 208, 216-217, *218,* 232
Cavalli-Sforza, L.L., **123-138**, 125, 129, 131-134, *136-138,* 139-145, 147, *227*
Cederlöf, R., 44, *52*
Chandler, J.A., 186, 188, *194*
Chen, S-Y, 103, *109*
Cheung, S. C-Y, 99-101, *109-110*
Childs, B., *119-122,* 105
Chilton, M.D., 97, *109*
Churchill, J.A., 45, *54*
Cicero, T.J., 99, *109*
Claiborne, R., 127, *137*
Clark, E.T., 35, *38,* 86, 88, *90*
Clarke, V.A., 72, *74*
Clayton, P.J., 130, *138*
Cohen, D., 44, *52*
Cohen, P.T.W., 103-104, *109*
Coleman, J.S., 61, *62*
Collins, R.L., 29, *30*
Condliffe, P., 189, *195*
Connor, J.D., 105, *110*
Cools, A.R., 115, *117*
Corah, N.L., 87, *90*
Costall, B., 114, *117*
Cowan, W.M., 99, *109*
Cowie, W., 228
Crow, J.F., **201-203**
Crowe, R.R., 145, 147, *149*
Curran, W.J., 181-184, 188, 191, *194*

Darwin, C., 142
Davies, W.E., 99, *109*
DeFries, J.C., 89, *90,* 145, *149,* 191, *195, 227*
Dejong, W., 37, *38*
Dekaban, A.S., 99, *108*
Denenberg, V.H., 33, *38*
Deol, M.S., 115, *118*
Dibble, E., 44, *52*
Dobzhansky, T., 77, 78, *83, 227*
Dofuku, R., 107, *109*
Donald, H.P., 140, *144*

Eaves, L.J., 11, *22,* 43, *52,* 80, *83*

---

*Page numbers in italic type indicate literature citations. Numbers in boldface type indicate contributions in this volume.

Eccles, J.C., 108, *109*
Eckland, B.K., **197-200**
Edwards, J.H., 45, *54,* 55, 56, *62,* 126, *138*
Ehrhardt, A.A., 107, *110*
Ehrman, L., *227*
Eichelman, B., 37, *38*
Eisdorfer, C., 35, *40*
Eleftheriou, B.E., 33, 37, *38, 39*
Elias, M.F., **33-40,** 37, *39*
Elias, P.K., 37n
Elston, R.C., 78, *83,* 129, *137*
Eppenberger, H.M., 100, *109*
Eppenberger, M., 100, *109*
Erdelyi, E., 131, *137*
Erlenmeyer-Kimling, L., **25-31,** 5, 9, *22,* 25, *30*
Ernst, A.M., 114, *118*
Etzioni, A., 199, 200
Evans, D.A.P., 120, *122*
Everly, J.L., 100, *110*
Eysenck, H.J., 141, 143, *144*

Falek, A., **179-195,** 86, *91,* 188-189, *194,* 197-201
Feit, H., 99, *109*
Feldman, M.W., 132-134, *137,* 142
Feldman, S.S., 125, *138*
Felsenstein, J., 174, 176
Fieve, R.R., 130, 137
Fisch, R.O., 107, *108,* 116, *117*
Fischer, K., 80
Fisher, M., 44
Fisher, R.A., 132, *137,* 142-143
Fleiss, M., 130, *137*
Fletcher, J.F., 185, *194*
Florini, J., 37
Ford, R.C., 121, *122*
Frankel, M.S., 189, *194*
Fraser, G.R., 174, *176*
Freeman, C.A., 69, *74*
Freeman, J.M., 131, *137*
Frenkenburg, W.K., 116, *118*
Freund, P.A., 183, *194*
Friberg, L., 44, *52*
Fulker, D.W., 10-13, 16, *23, 24,* 47, 51, *53,* 80, 83
Fuller, J.L., 15, *22,* 29, *30,* 37, *39,* 227-228

Galton, F., 132
Gehring, M., 99, *110*
Geils, H., 37, *39*
Gershovitz, H., 44, *53*
Gesell, A., 88, *90*
Giblett, E.R., 103, 109
Gibson, D.G., 38, *39*
Gibson, J.B., 69, 72, *74*
Ginsburg, B.E., 25, *30*
Glass, D.C., *227*

Gluck, J.P., 34, *39*
Goldstein, A.D., 116, *118*
Goldstein, O., 190, *194*
Goodwin, D.W., 49, *54,* 146, 148, *149*
Gottesman, I.I., **221-227,** 7, *22,* 25, *30,* 223, *225,* 227
Gould, S.J., 87, *90*
Grawe, J., 44, *52*
Green, H.C., 61, *62*
Green, M.C., 37, *39*
Gribbin, K., 205, *219*
Grossfield, R.M., 99, *109*
Grouse, L., 97, 98, *109*
Guilford, J.P., 16, *22*
Guillery, R.W., 115, *118*
Gurd, R.D., 99, *109*
Gustafson, J.M., 186, *195*
Guttman, R., 87, *90*
Guze, S.F., 146, 148, *149*

Hahn, W.E., 97, *110*
Haldane, J.B.S., 21, *22,* 143, *144*
Hamberger, A., 100, *110*
Hamilton, M., 189, *194*
Hammurabi, 180
Hancock, C., 69, *74*
Harlow, H.F., 34, *39*
Harpring, E.B., 14, *24*
Harris, H., 102, *110,* 119, *122*
Harris, M., 189, *195*
Harrison, G.A., **65-75,** 65, 69, 73, *74, 75,* 77, 78, 85, 86
Hartlage, L.C., 50, *53,* 87, *90*
Harvald, B., 44
Haseman, J.K., 78, *83*
Hauge, M., 44, *53*
Henderson, N.D., **5-24,** 7, 8, 14, 20, *22, 23,* 25, 26, *30,* 33-35, 87, 98
Hermanson, L., 146, 148, *149*
Heston, L.L., 145-146, *149,* 222-223, *225*
Hill, M.S., 50, *53*
Hill, R.N., 50, *53*
Hilton, B., 189, *195*
Hiorns, R.W., 69, 72, *74, 75*
Hippocrates, 180
Hirsch, J., *227*
Hirschhorn, K., 193, *195*
Hobson, C.J., 61, *62*
Hofstadter, R., 142, *144*
Hopkinson, D.A., 119, *122*
Horn, J.L., 18, *23,* 47
Howard, M., 47, *52*
Hoyer, B.H., 97, *110*
Humphreys, L.G., 17, *23*
Hunt, J. McV., 6, *23*

Insel, P., 50, *53*

Jablon, S., 44, *53*
James, J.W., 129, *137*
Jarvik, L.F., **85-91,** 5, 15, *22, 23,* 35, *38,* 86, 88, 89, *90*
Jayakar, S.D., 130, *137*
Jencks, C., 47, *53*
Jensen, A.R., 13, 19, *23,* 126, 135, *137,* 216, *218*
Jinks, J.L., 10-13, *23,* 47, 51, *53,* 80, *83*
Job, T.H., 113, *118*
Johnson, G.B., 116, *118*
Johnstone, E.C., 120, *122*
Jonsson, E., 44, *52*

Kaas, J.H., 115, *118*
Kaij, L., 44, *52*
Kallman, F., 85, 86, *91,* 171
Keele, D.K., 105, *110*
Keeves, J.P., 61, *62*
Kety, S.S., 145, *149,* 190
Kidd, K.K., 129, *137*
King, R.C., 174, *176*
Kohn, M.L., 28, *30*
Kolakowski, D., 50, 52, *87,* 90
Kopp, G.A., 61, *62*
Kristy, N.F., 51, *52*
Küchemann, C.F., 69, *74, 75*
Kuse, A.R., 89, *90*

Ladimer, I., 179, 181, 183, *195*
Laird, C.D., 97, *110*
Lappe, M., 186, *195*
Lashley, K.S., 18, *23*
Layzer, D., 141, *144*
Lerner, I.M., **139-144,** 145
Lindzey, G., 37, *39*
Loehlin, J.C., **41-54,** 10, 18, 23, 45, 51, 52, *53,* 55, 57, 60, *62*
Lorge, I., 86, *91*

McCall, R.B., 126, *137*
McCarthy, G.J., 97, 98, *109-110*
McClearn, G.E., **1-3,** 89, *90,* 115, 145, *149,* 191, *195, 227*
McKeown, T., 45, *54,* 55, 56, *62,* 126, *137-138*
McKusick, V.A., 127, *137*
McPartland, J., 61, *62*
Macbeth, H.M., 69, *74*
Mahler, H.R., 99, *109*
Manosevitz, M., 37, *39*
Marjoribanks, K.M., 61, *62*
Marks, J.F., 105, *110*
Marolla, F.A., 19, *21,* 55, 57, *62,* 127, *136, 138*
Marsden, C.D., 114, *118*
Marsh, W., 120, *122*
Matthysse, S., 114, *118*
Mayr, E., 78, *83*
Mendel, G., 142

Mendlewicz, J., 130, *137*
Meredith, W., 52, *53,* 86, 87
Messeri, P., 33, *39*
Meyer, G., 34, *39*
Money, J., **151-170,** 87, *91,* 107, *110,* 121, 165, 168-169, *170,* 170-173, 176
Mood, A.M., 61, *62*
Moor, L., 55, 56, *62*
Moore, B.W., 99, *109-110*
Moore, W.J., 99, *109*
Morgenroth, V.H. III, 113, *118*
Mosychuk, H., 61, *62*
Motulsky, A.G., 100-101, 103-104, *109-110,* 131, *138,* 174, *176,* 189, 192, *194-195*
Mudd, S.H., 131, *137*
Murray, R.G., 192, *195*
Myrianthopoulos, N.C., 45, *53*

Nakata, M., 60, *62*
Nance, W.E., 60, *62*
Naylor, R.J., 114, *117*
Neel, J.V., 44, 47, *53, 54*
Nesselroade, J.R., 205, *218*
Newman, R.W., 183, *195*
Nichols, P.L., 45, 51, *53*
Nichols, R.C., 44, 45, *53, 54*
Nirenberg, M., 108, *108*
Nyhan, W.C., 105, *110*

O'Donnell, T.J., 185, *195*
Ohno, S., 107, *109*
Oliverio, A., 33, *38, 39*
Olson, C.O., 116, *118*
Omenn, G.S., **93-111,** 27, *30,* 97-100, 103-104, 107, *109-110,* 113-117, 119-120, 131, *138, 227*

Packman, P.M., 100, *110*
Palay, S.L., 96, *111*
Pare, C.M.B., 120, *122*
Paul, T.D., 60, *62*
Penhoet, E., 101, *110*
Perez, V.J., 99, *110*
Pettigrew, K.D., 9, *21*
Piaget, J., 20, 78, 82, *83,* 127-128
Pick, A., 79, *83*
Pickel, V.M., 113, *118*
Pignatti, P.F., 131, *138*
Pollin, W., 44, *52*
Potter, R.K., 61, *62*
Purcival, T., 180

Quarles, R.H., 100, *110*

Rajkumer, T., 101, *110*
Record, R.G., 45, *54,* 55, 56, *62,* 126, *137-138*
Reed, S.C., **171-172**
Reich, T., 130, *138*

Reinert, G., 11, *21*
Reis, D.J., 113, *118*
Richelson, E., 108, *108*
Richterick, R., 100, *109*
Riegel, K.F., 34, 35, *39*
Riegel, R.M., 34, 35, *39*
Robitscher, J., 183, 189, *195*
Roblin, R., 186, *195*
Romano, J., 180, *195*
Rose, S.F., 186, 188, *194*
Rosenthal, D., 28, *31,* 145, *149,* 227
Roth, R.H., 113, *118*
Russell, E.S., 37, *39*
Rutter, W.T., 101, *110*

Saatcioglu, A., 69, *74*
Saenger, G., 127, *138*
Sander, G., 85, 86, *91*
Santachiara, S.A., 131, *137*
Scarr-Salapatek, S., **77-83,** 9, *23,* 45, *54,* 82, *83*
Schaie, K.W., **vii-ix, 205-219,** 11, *23,* 34, 35, 38, *39,* 82, *83,* 205-209, *218-219*
Schlager, G., 37, 38, *39*
Schuckit, M.A., 49, *54*
Schull, W.J., 47, *54*
Schulsinger, F., 145-146, 148, *149*
Scott, J.P., *228*
Sears, R.R., 125, *138*
Shah, S.A., 189, *194*
Shaw, A., 185, *195*
Shields, J., 12, 13, *23,* 46, *54,* 227
Shneour, E.A., 140, *144*
Shockley, W., 198
Shooter, E.M., 96, 99, *109-110*
Sidman, R.L., 37, *39*
Siegel, F.S., 107, *108,* 116, *117-118*
Siegel, S., 171
Siekewitz, P., 189, *195*
Simonson, E., 35, *39*
Skeels, H.M., 126, *138*
Skodak, M., 126, *138*
Slater, E., *228*
Smilek, P.G., 114, *118*
Smith, R.T., 60, *62*
Snook, A., 148
Sorrentino, R., 37n
Sotelo, C., 96, *111*
Spearman, C., 19, 52
Spencer, H., 142
Spieth, W., 35, *40*
Sprott, R.L., 33, 37, *39, 40*
Stafford, R.E., 50, *54,* 87, *91*
Stanley, J.C., 205, *218*
Stein, Z., 127, *138*
Stern, S.E., 9, *22*
Stice, G.F., 51, *52*
Streeten, D., 37n

Suntzeff, V., 99, *109*
Suomi, S.J., 34, *39*
Susser, M., 127, *138*

Taylor, C.C., 60, *63*
Taylor, E., 142
Tellegen, A., 107, *108*
Tettenborn, U., 107, *109*
Thiessen, D.D., 33, 34, 37, *39,* 40
Thoday, J.M., 72, *75*
Thompson, H., 88, *90*
Thompson, W.R., 15, *22, 23,* 227
Thurstone, L.L., 16, *23*
Tizard, B., 135, *138*

Ungerstedt, U., 113-114, *118*

Vale, C.A., 25, *31*
Vale, J.R., 25, *31*
Vallot, F., 55, 59, *63*
Vandenberg, S.G., 18, *23,* 52, *53,* 89, *90,* 228
Varma, A.O., 15, *23,* 89, *91*
Vernon, P.E., 17, *23*
Von Bracken, H., 60, *63*

Waddington, C.H., 78, *83,* 222, *225*
Wang, L., 131, *137*
Warren, E., 183
Weber, G., 105, *111*
Weinfeld, F.D., 61, *62*
Welner, J., 145, *149*
Wender, P.H., 145, *149*
Werner, H., 78, *83*
White, T.A., 120, *122*
Wigley, J.M., 69, *74*
Wilcock, J., 16, *24,* 33, 34, *40*
Wilde, G.J.S., 46, *54*
Wilkie, F., 35, *40*
Willerman, L., 45, *54*
Williams, R., 141
Williams, R.B., 37, *38*
Williams, T., 61, *63*
Wilson, J.R., 89, *90*
Wilson, R.S., 14, *24,* 126, *138*
Winokur, G., 49, *54,* 130, *138,* 146, 148, *149*
Wirt, R.D., 107, *108,* 116, *117*
Witkop, C.J., 115, *118*
Wohlwill, J.R., 205, *219*
Wolf, A., 135, *138*
Wolf, R.M., 61, *63*
Wright, S., 21, *24*

York, R.L., 61, *62*
Young, M.K., Jr., 141, *144*
Yu, P.L., 60, *62*

Zatz, M., 100, *111*
Zazzo, R., 60, *63*

Subject Index

Abortion, 48, 127, 148, 161, 185, 192-193, 202
Additive effects, 6, 8, 10, 12, 43, 47-48, 65, 98, 103
Adoption studies, 47-48, 51, 80-82, 126, 134-135, 145-148
Adrenal cortex, 104
Adrenogenital syndrome, 107, 229
Aging, 15, 37, 115, 205-206
   age changes, 88, 206-208, 209-217
   age differences, 14, 18, 50, 107, 206-208, 211, 213
   age of mother, 56-57
Albinism, 115, 229
Alcoholism, 27, 48-49, 146-148
Amniocentesis, 127, 161, 174, 185, 192, 198, 202
Animals as subjects. *See* Models, animal; *see also* species name of interest
Antisocial behavior, 187
   criminality, 27, 145-147
Artificial insemination, 191-192, 202
Assay techniques. *See also* Pharmacogenetic analysis
   chromatography, 99, 229
   electrophoresis, 94, 99-102, 104, 119, 131, 230
Assortative mating, 5, 10, 12, 50, 72, 78, 80, 126, 191, 205, 215, 216
   coefficient, 80

Barr body, 154, 229
Bayley scales of mental and motor development, 14-15
Birth control, 162-163, 166, 170, 175, 193, 198
Birth order, 19, 57-59
Black race, 9, 78, 135, 198
Blending theory of inheritance, 132-133, 142
Blood grouping, 44, 71, 102
   and intelligence, 72
Blood pressure. *See* Cardiovascular disease
Brain, gene action in, 96-105, 113-117, 120, 130-131. *See also* Enzymes
   cortex, 97, 99
Brain damage, 18, 27, 127. *See also* Mental retardation; Phenylketonuria (PKU)
Brain weight in mice, 8

Caenorhabditis elegans, 114
California Psychological Invenotory, 45
Cardiovascular disease, 35-38, 89
Carrier, 151-153, 161, 164, 172-174, 229
   carrier detection, 176, 192
   sexual attitudes and behavior of, 162-166

Choreoathetosis, 105, 229
Civil rights. *See* Ethical issues; Human subjects; Informed consent
Cloning, 98, 108, 193
Cohort-sequential method, 207, 211, 214, 216-217
Cohort effects, 206-209, 211-217
Color vision, 65, 105, 169
Conditionability, 12. *See also* Learning
Conditioning, 16, 76. *See also* Learning
Congenic lines, 33, 37, 230
Cortical evoked potentials, 115
Contrast effect in twins, 46
Co-twin control method, 88-89, 202, 230
Cretinism, 107, 230
Criminality. *See* Antisocial behavior
Critical loss, 88-89
Critical period, 29
Cross-cultural research, 18, 65, 79
Cross-sequential method, 208, 213, 215, 217
Cultural evolution, 131-136, 139, 142, 147
Cyclic AMP, 108
Cystathionenuria, 106
Cytogenetic diagnosis, 167-168

Deafness, 27, 115
Demographic structure, 67-69, 78. *See also* Social class structure; Social mobility
Depression, 120. *See also* Psychoses
   as a result of genetic shock, 151
Developmental changes, 216. *See also* Aging, age changes
Developmental lag, 107
Diallel cross, 6-7, 14
Diet. *See* Nutrition
DNA (deoxyribonucleic acid), 89, 93-94, 96-99, 102, 230
   unique sequence, 97-98
DNA-RNA hybridization, 96-98, 102, 119
Dominance, 6, 8, 10, 16, 43, 47-48, 50, 151, 153, 174, 230
Dominoes Test, 12-13
Down's syndrome, 27, 127, 160, 192-193. *See also* Mental retardation
Drug action, 114, 120, 130-131

Ecology
   ecological situation, 66
   ecological validity, 81
Endocrine activity, 2, 87, 93-94, 104, 107
Environmental complexity, 7-8
Enzymes, 2, 94-95, 100-106, 108, 113-116, 120-121, 130-131, 193, 223-224

acetyl transferase, 120
acid phosphatase, 102-103
aldolase, 101
creatine phosphokinase, 100
cystathionine synthetase, 106
dopa decarboxylase, 104, 131
dopamine beta-hydroxylase, 101, 104, 113-115
enolase, 103
5 hydroxy-tryptophan, 113, 131
hexokinase, 103
hydroxysteroid dehydrogenase, 101
hypoxanthine-guanine phosphoribosyl-transferase, 105
lactase, 223-224
lactate dehydrogenase, 103
malic enzyme, 103-104
monoamine oxidase, 104, 120
phenylalanine hydroxylase, 121
phoglycerate kinase, 103
phosphoglycerate mutase, 100-101
tetrahydrofolic reductase, 131
tyrosine hydroxylase, 104, 113
Eosinophilia, 171, 230
Equality of Educational Opportunity report, 61
Erythroblastosis fetalis, 190
Ethical issues. *See also* Human subjects
  behavioral genetics research, 123-125, 136, 143, 191-193, 198-202
  genetic counseling, 161, 164, 168, 173-175
Eugenics, 129, 180, 189, 191-192, 197, 199, 230
Euphenics, 176, 230
Euthenics, 190-191, 230
Evolution, 66-67, 77-79, 82, 97, 103, 131-133, 142, 212, 222-223. *See also* Cultural evolution
Eysenck Personality Inventory, 71

Factor analysis, 16-18, 52
Family studies, 67, 69, 72-73, 77, 79-82, 85. *See also* Siblings
Fertility patterns, 67, 70, 78
  and IQ, 197
Fetal brain proteins, 99-101
Fluid-crystallized intelligence, 17-19, 87
French National Demographic Institute, 55
Full sibs. *See* siblings

Galactosemia, 127, 223, 230
Gene-environment correlation, 5, 10-14, 19, 26, 28, 42-43, 80, 135, 143
Gene-environmental interaction, 5-15, 19, 25-28, 30, 37-38, 42, 81-82, 87, 135, 139, 143
Genetic counseling, 2, 55, 151-176, 192, 201

Genetic drift, 132, 231
Genetic shock, 151
Guinea pig, 99

Half sibs. *See* siblings
Handedness, 45, 71, 79
Height, 87
Hemoglobin diseases, 128
Hemophilia, 153
Heritability estimates, 9, 15, 17, 21, 128-129, 131, 135-136, 140-141, 191, 205, 208, 215-217, 231
Histidinemia, 105, 231
Holland Vocational Preference Inventory, 45
Homocystinuria, 105-106, 231
Human subjects, 38, 136, 179-188, 201. *See also* Ethical issues; Informed consent
Huntington's chorea, 27, 174, 188, 192, 231
Hybrids, 73-74, 85-86, 231. *See also* DNA-RNA hybridization
  somatic cell hybrids, 130
Hypertension. *See* Cardiovascular disease
Hyperuricemia, 105, 231

Impulsivity, 41-42
Informed consent, 181, 183-187, 193, 198. *See also* Human subjects
Intelligence, 2, 5-6, 10, 12-21, 26-27, 35, 43-45, 47, 50-52, 55-57, 72-73, 78, 86-88, 107, 120-121, 124-129, 135, 140-141, 143, 169, 191, 197, 203. *See also* Spatial relations
  cognitive abilities, 60, 65
  cognitive factors, 89
  memory, 99
Introversion-extroversion, 12, 72
IQ. *See* intelligence
Isozymes, 100, 104, 232

Karyotype, 152, 155-160, 186-187, 191, 232
Kibbuzim, 81
Klinefelter's syndrome, 159, 168-169, 188, 193, 232

Lactose tolerance test, 223-224
Language, 79, 82, 99, 107, 133
  speech defect, 105
Learning, 6, 15-16
  difficulties, 107, 120, 168-169
  instruction methods, 88-89, 136
Lesch-Nyhan syndrome, 105, 114, 127, 232
Life-span
  behavior genetics, 34-35, 37, 82, 125
  of investigators, 85, 125, 139-140
Limited resource model, 11
Longitudinal methods, 11, 29-30, 34, 82, 85, 125-126, 139-140, 205-208. *See also*

Cohort-sequential method; Cross-
sequential method
combined with cross-sectional methods,
11, 34, 82

Manic-depressive psychosis, 27, 130, 190
Markers, 72, 130
Maternal effects, 6, 48, 80, 232
Maudsley Personality Inventory, 12
MAVA. *See* Multiple abstract variance
analysis
Membrane receptor functions, 94-95
Mental illness, 189, 191, 197. *See also*
Psychopathology
Mental retardation, 2, 27, 94, 105-107,
120-121, 127, 175, 189, 191, 197. *See
also* Klinefelter's syndrome; Phenyl-
ketonuria (PKU)
Metachromatic leukodystrophy, 105-106,
232
Mice. *See* Rodents
Migration, 212. *See also* Social mobility
Minnesota Multiphasic Personality Inven-
tory (MMPI), 47, 50
Models, 41, 43, 49, 68, 78, 142
animal, 34-38
computer simulation, 95
Monkeys, 98, 102
Mortality patterns, 70, 88-89
Motion sickness, 60
Motor development in children, 65
Multiple abstract variance analysis (MAVA),
10, 18, 51, 208, 232
Multivariate analysis, 10-11, 51-52, 61. *See
also* Factor analysis; Multiple ab-
stract variance analysis

National Merit Scholarship Qualifying Test,
45-46, 55, 58
Neuroticism, 12. *See also* Personality
in family of genetically defective children,
151, 162, 166-167
Neurotransmitters, 101, 104, 106, 108,
113-115, 120, 131
Nutrition, 7, 18, 27, 43, 89, 116, 126-127, 134,
140, 202

Otis intelligence scale, 13
Ouabain, 96

Peer review, 182, 200
Personality, 12-14, 19-20, 46, 49-50, 60, 71,
136, 175
Pharmacogenetic analysis, 104, 117
Phenylalanine tolerance test, 121, 233
Phenylketonuria (PKU), 27, 105, 116-117,
121, 124, 127, 129, 176, 189, 233

Piagetian tests, 127-128
Plasticity, 6, 20, 78, 81, 133-134
Population structure. *See* Demographic
structure
Proteins, 93-94, 99-103. *See also* Enzymes
synthesis, 96-97
Psychopathology, 26-29, 103, 106, 123-124,
167, 169, 175. *See also* Alcoholism;
Antisocial behavior; Mental illness,
Mental retardation; Personality;
Psychoses
Psychoses, 114, 130. *See also* Depression;
Manic-depressive psychosis;
Schizophrenia
PTC (phenylthiocarbamide), 65, 233

Rabbit, 99
Race, 73, 233. *See also* Black race
and IQ, 124-125
Rats. *See* rodents
Raven Progressive Matrices Test, 13, 19, 55
Reaction range, 7-8, 25, 27
Recombinant-inbred strains, 33, 37, 233
RNA (ribonucleic acid), 93-94, 96-99, 102,
119, 234
Rodents
mice, 6, 7-9, 14, 27, 36-38, 97-99, 102, 108,
113-115
rats, 113
shaker-waltzer type mutants, 115
varitint-waddler mice, 115
Rural populations, 67-73, 86

Schizophrenia, 2, 26-28, 30, 61, 106, 114,
128-129, 141, 145-146, 190-191. *See
also* Psychoses
Seizures, 27, 29
Sequential methods. *See* Cohort-sequential
method; Cross-sequential method;
Time-sequential method
Sex chromosome-related effects, 6, 50,
55-56, 73, 87-88, 95, 104-105, 130, 139,
153, 174, 234. *See also* Klinefelter's
syndrome; Turner's syndrome; XYY
syndrome
Sex differences, 72, 87-88, 104
Siblings. *See also* Family studies
full, 10-11, 43, 49-51, 60, 79, 116, 230
half, 48-49, 60, 79, 202, 231
Sickle-cell anemia, 176, 180, 234
Single gene effect, 2, 27, 33-34, 128
Sleep behavior, 61
Social class structure, 2, 28, 66-67, 69-72
Social implications, 123-125, 128-129, 143.
See also Ethical issues
Social mobility, 67-72, 186
Socioeconomic status, 47, 56, 58, 81. *See*

also Gene-environment correlation; Gene-environmental interaction; Social class structure
Spatial relations, 50, 52, 87, 107
Sporadic heredity, 152, 170
Stanford-Binet, 51
Sterilization, 166, 170, 193
   in Nazi Germany, 179
Stress, 28-29, 36-37, 44, 94
Sulfatide, 106, 234

Tay-Sachs disease, 127, 161, 175, 234
Testicular feminization syndrome, 107, 234
Thalidomide, 181-182
Time-of-measurement effects, 15, 29, 41, 206-208, 209-217
Time-lag method, 206-207, 211-213
Time-sequential method, 208, 212, 214, 217
Turner's syndrome, 157, 168-169, 193, 234

Twins, 1, 9-13, 15, 18, 43-47, 51, 55, 60, 79-81, 86, 88-89, 126, 135-136, 140, 202, 205, 209, 216
   as parents, 60, 79-80, 202
   zygosity diagnosis, 44-45

Urban populations, 67-33, 86

Voice prints, 61

Wechsler intelligence tests, 51, 61, 71
Wilson's disease, 190

X chromosome. *See* Sex chromosome-related effects
XXY. *See* Klinefelter's syndrome
XYY syndrome, 158, 168-169, 175, 186-188, 191, 193

Invited Workshop Participants

Dr. Paul Baker
Pennsylvania State University

Dr. L. L. Cavalli-Sforza
Stanford University

Dr. Barton Childs
Johns Hopkins University

Dr. James F. Crow
University of Wisconsin

Dr. John C. DeFries
University of Colorado

Dr. Bruce K. Eckland
University of North Carolina

Dr. Merrill F. Elias
Syracuse University

Dr. L. Erlenmeyer-Kimling
New York State Psychiatric Institute

Dr. Arthur Falek
Georgia Mental Health Institute

Dr. Irving Gottesman
University of Minnesota

Dr. G. Ainsworth Harrison
Oxford University

Dr. Norman D. Henderson
Oberlin College

Dr. Lissy F. Jarvik
University of California at Los Angeles

Dr. I. Michael Lerner
University of California at Berkeley

Dr. John C. Loehlin
University of Texas

Dr. William Meredith
University of California at Berkeley

Dr. Robert F. Murray
Howard University

Dr. Gilbert S. Omenn
University of Washington

Dr. Sheldon C. Reed
University of Minnesota

Dr. Sandra Scarr-Salapatek
University of Minnesota

Dr. W. R. Thompson
Queens University (Kingston, Ontario)

Dr. Stephen G. Vandenberg
University of Colorado

Members of the Developmental Behavioral Sciences Study Section National Institutes of Health

April 1974

Chairman:

Dr. K. Warner Schaie
Department of Psychology and
Andrus Gerontology Center
University of Southern California

    Dr. V. Elving Anderson
        Dight Institute of Human
           Genetics
        University of Minnesota

    Dr. Ernest Q. Campbell
        Department of Sociology
        Vanderbilt University

    Dr. Frances M. Carp
        The Wright Institute
        Berkeley, California

    Dr. Rue L. Cromwell
        Department of Psychiatry
        University of Rochester

    Dr. Edwin D. Driver
        Department of Sociology
        University of Massachusetts

    Dr. Donald K. Freedheim
        Department of Psychology
        Case Western Reserve
           University

    Dr. Goldine C. Gleser
        Department of Psychiatry
        University of Cincinatti

Dr. Phillip B. Gough
  Department of Psychology
  University of Texas

Dr. Janellen Huttenlocher
  Department of Psychology
    and Education
  Columbia University

Dr. Hilda Knobloch
  Department of Pediatrics
  Albany Medical College

Dr. Alvin M. Liberman
  Department of Psychology
  University of Connecticut

Dr. Gerald E. McClearn
  Institute of Behavioral Genetics
  University of Colorado

Dr. John Money
  Department of Psychiatry and
    Behavioral Sciences and
  Department of Pediatrics
  The Johns Hopkins Hospital

Dr. Gilbert F. Young
  Departments of Neurology
    and Pediatrics
  Medical College of South
    Carolina

Executive Secretary:

Dr. Bertie H. Woolf
Division of Research Grants
National Institutes of Health

About the Editors

K. Warner Schaie received the B.A. from the University of California at Berkeley and the M.S. and Ph.D. from the University of Washington at Seattle. He is Professor of Psychology and Associate Director for Research at the Andrus Gerontology Center of the University of Southern California at Los Angeles. A past president of the Division of Adult Development and Aging of the American Psychological Association, Dr. Schaie has taught at the University of Nebraska-Lincoln and West Virginia University, where he was chairman of the psychology department, and directed a life-span developmental psychology training program. His research interests are concerned with changes in intellectual abilities and personality over the adult life span. As part of these interests he has been involved in the development of strategies to segregate ontogenetic age change from sociocultural input in studies of human development. Dr. Schaie's publications include *Color and Personality* (1964), *Theory and Method of Research on Aging* (1968), *Life-Span Developmental Psychology: Personality and Socialization* (1973), and *The Handbook of the Psychology of Aging* (1976).

V. Elving Anderson received the Ph.D. in zoology from the University of Minnesota in 1953. He is Assistant Director of the Dight Institute for Human Genetics and Professor of Genetics and Cell Biology at the University of Minnesota. He has also taught at Bethel College and has served with the National Institute of Neurological Diseases and Blindness, first as a visiting scientist and then as a consultant. Dr. Anderson is Chairman of the Developmental Behavioral Sciences Study Section of the National Institutes of Health (1974-75) and has served as secretary of the Behavior Genetics Association. He is the author or coauthor of numerous articles that have been published in professional journals and scholarly works; his other publications include *The Psychoses: Family Studies* (1973), which he coauthored with S. C. Reed, C. Hartley, V. P. Phillips, and Johnson Nelson.

Gerald E. McClearn received the Ph.D. in psychology from the University of Wisconsin in 1954. Since 1966 he has been Professor of Psychology and Director of the Institute for Behavioral Genetics at the University of Colorado in Boulder. Previously he was a faculty member at the University of California, Berkeley, and a NAS-NRC Senior Postdoctoral Fellow in Physiological Psychology at the Institute of Animal Genetics in Edinburgh. President of the Behavior Genetics Association during 1974, Dr. McClearn has served on several NIH/NIMH review committees and panels and has been a member of the Social Science Research Council Committee on Biological Bases of Social Behavior since 1961. His research on mouse behavioral genetics has emphasized the behavioral domains of alcohol preference and sensitivity to the effects of alcohol. His human research is concerned with the genetics of cognition and perceptual processes, including dyslexia. Dr. McClearn's publications include numerous articles that have appeared in professional journals and scholarly books, and *An Introduction to Behavioral Genetics* (coedited with J. C. DeFries), published in 1973.

John Money was educated at the University of New Zealand and received the Ph.D. from Harvard University in 1952. In 1951 he began working as a medical psychologist in the Pediatric Endocrine Clinic and Department of Psychiatry at The Johns Hopkins University and Hospital, where he is now Professor of Medical Psychology and Associate Professor of Pediatrics. He is also director of the Psychohormonal Research Unit, where he has pioneered human clinical studies in behavioral endocrinology, behavioral cytogenetics, and behavioral sexology. Dr. Money is internationally known for publications on gender identity and the origins and options in sex differences.